NATURE'S
GIANTS

NATURE'S GIANTS

The Biology and Evolution of the World's Largest Lifeforms

Graeme D. Ruxton
Foreword by Norman Owen-Smith

Yale UNIVERSITY PRESS
New Haven and London

Yale University Press books may be purchased in quantity for educational, business, or promotional use. For information, please e-mail sales.press@yale.edu (U.S. office) or sales@yaleup.co.uk (U.K. office).

This book has been composed in Gotham and Gotham Narrow. Printed on acid-free paper.

Conceived and designed by The Bright Press, part of the Quarto Group, Level 1, Ovest House, 58 West Street, Brighton, UK BN1 2RA

Designed by Luke Herriott.

Project managed by D & N Publishing, Wiltshire.

Commissioned by Jacqui Sayers.

Printed in China by 1010 Printing International Ltd.

Library of Congress Control Number: 2018962449

ISBN 978-0-300-23988-1 (hardcover : alk. paper)

A catalogue record for this book is available from the British Library.

10 9 8 7 6 5 4 3 2 1

Contents

Foreword by Norman Owen-Smith   9

Introduction   11

1   Life on a Large Scale   12

2   Dinosaurs   22

3   Massive Mammals   38

4   Giants of the Deep   70

5   Giants of the Skies   100

6 Giant Insects 124

7 Immense Invertebrates 142

8 Record Reptiles and Amphibians 170

9 Green Giants 202

Final Thoughts 218

Further Reading 220

Index 220

Acknowledgments 224

Foreword

Humans have an enduring fascination with very large animals, as seen in examples of rock art across Europe and Africa. Sadly, prehistoric megaherbivores, or plant-eating animals weighing more than 2,200 lb (1,000 kg), disappeared everywhere outside Africa and tropical Asia soon after modern humans appeared on their continents. Today, those megaherbivores that have survived remain under threat from sustained hunting pressure from humans—something I highlighted during my study of one of these magnificent creatures, the white rhino (*Ceratotherium simum*).

In this book, Professor Ruxton explains what limits the sizes reached by the largest organisms. He weaves together a fascinating account of adaptive morphology and physiology to interpret factors governing the body sizes attained by different life-forms, especially those that prevent them from growing any larger. Importantly, he also addresses conservation implications, because it is some of the largest species—including the white rhino—that are most at risk from human population growth.

This exceptionally readable account expands on concepts of what large size means for organisms, from big plant-eating mammals to a wide range of other taxa. It explains why the largest living mammals are not elephants, but whales, which are not herbivores but instead feed on tiny crustaceans called krill. And why some of the reptilian herbivores alive during the Jurassic and Cretaceous were significantly larger than any terrestrial mammal, but living alongside them were numerous much smaller reptiles. It also looks at birds, and asks why none have approached the size of the biggest mammalian herbivores even though they have descended from a group of dinosaurs. And it explores why giant spiders, despite the phobias they generate among humans, do not grow large enough to cause us any real harm. I trust that you, the reader, will enjoy this fascinating exploration of nature's giants as much as I did.

Norman Owen-Smith
Research Professor Emeritus
University of the Witwatersrand, Johannesburg, South Africa

◀ It has been claimed that hippos (*Hippopotamus amphibius*) kill more people than any other African mammal, including lions (*Panthera leo*).

Introduction

Size Matters

Size really does matter when it comes to your experience of the world around you: the puddle you step over without thinking about it will be all of the world the thousands of microscopic rotifers that live there ever know. Size affects every aspect of life—to give just a couple of examples, bigger things live longer but leave fewer offspring, while a group of mice weighing 500 lb need to eat more food, drink more water, and breathe more oxygen than a single 500 lb cow. We take it for granted that animals of a certain type are a certain size, but why is this the case? Why aren't there spiders big enough to eat us alive, or whales small enough to keep in the bath? Exploring these questions helps us understand more clearly how all sorts of natural organisms function, and by studying the biggest or the smallest, the underlying drivers of size should become clearer. This book focuses on extremely large organisms because they have been studied in more depth than those that are extremely small—and also because they are awe-inspiring and often downright terrifying!

How this Book is Organized

The next 200 pages are arranged into nine chapters. For the most part I look at specific living and extinct giants, and organize them into rough categories in Chapters 2 to 9. I apologize now if a particular favorite of yours gets only a passing reference or no reference at all—I have only so many pages to play with and some tough choices had to be made. As we shall see in Chapter 2, most dinosaurs were either big or really big and more than 700 species have been named, so I could easily have filled the book with these alone. Instead, I have chosen to consider only a select few of the really big dinosaurs so that we can also look at the giants of a range of other animal and plant groups.

Chapter 1 is a little different, in that it is the only one that doesn't focus on specific species. Rather, it overviews some general themes that we will meet throughout the book, and saves me repeating myself when we meet a similar issue affecting the giants of different groups.

I felt I should finish with a final spread trying to explain why most organisms aren't huge, and why one species that has a big body but isn't especially large—namely *Homo sapiens*—has dominated the world to a degree unmatched by any giant.

◀ Generally just a little heavier than the lion (*Panthera leo*), the tiger (*Panthera tigris*) is the biggest of the big cats.

Chapter 1
LIFE ON A LARGE SCALE

I have arranged this book with each chapter looking at a different broad type of organism. However, there are some general principles that pervade the whole book, and things will make a lot more sense if we cover these first. A few of these ideas will be a refresher of some very simple ideas from geometry and physics that you might remember from school. Then we move on to some theories concerning evolution and how populations adapt to the environment around them, and ideas from ecology about how populations of different individuals fit together to form an ecosystem.

Volume, Surface Area, Mass, and Metabolism

Imagine an animal that stays the same in every respect except that it is scaled upwards in size. Its mass, metabolism, and surface area all increase as it gets bigger, but crucially they do so at different rates. This has huge consequences for the biggest animals, so we need to spend a little time looking more closely at this phenomenon.

Some Simple Scaling

Imagine two cubes as shown to the right. Cube A is 1 ft long on every side and cube B has sides that are twice as long (2 ft). The volume of cube A can be calculated as its length multiplied by its breadth multiplied by its height (1 ft × 1 ft × 1 ft = 1 ft³). The surface area of one side of cube A is 1 ft × 1 ft = 1 ft², and as it has six identical sides, its total surface area is 6 ft². From this, it is easy to see that the volume of cube B is 8 ft³ and its total surface area is 24 ft². So, as we double the length of a cube (going from cube A to cube B), we increase its volume eightfold. If the two cubes are made of the same material, and drawing on the principle that mass is proportional to volume, then the mass of cube B also increases by a factor of eight. In comparison, the total surface area increases by a factor of only four. So, both surface area and mass increase with increasing size, but mass increases proportionally faster than surface area. This isn't just true of cubes. If we take a sphere and double its diameter, we increase its mass eightfold and its surface area only fourfold. In fact, this principle is true for any shape. But why does it matter?

Instead of considering a cube or a sphere, imagine a bird. As with the cube, if we double the size of the bird, its weight (which is the force of gravity acting on its mass) will increase eightfold but its surface area will increase only fourfold. A bird in flight has to overcome the force of gravity bearing down on it with the lift generated by its wings. The force of gravity acting on it is directly proportional to its weight, with a bird twice as long from nose to tail experiencing eight times the downward force of gravity. The upward lift generated by the bird's wings depends on how forcefully it beats them downwards and on their surface area. Using our simple scaling, the bird that is twice as long will have a wing area that is four times the wing area of the original bird. If it beats its wings with the same force

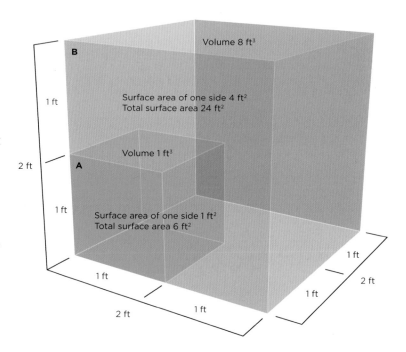

as the smaller bird, it will fall like a stone, so somehow it must find the extra muscle power to beat its wings twice as forcefully. From this, it is clear that the bigger a bird gets, the harder it has to work in order to stay airborne—until eventually there will be a limiting size, where the bird simply can't beat its wings hard enough to overcome the force of gravity. This limit on the size a bird can be and still fly stems directly from the way surface area and mass scale differently with increasing size.

Declining Muscle Power

In fact, the situation is even worse for the large flying bird. Imagine that both the small bird and the big bird have the same anatomy, and in each 40 percent of their volume is devoted to flight muscles for beating their wings. You might reasonably think that the larger

◀ Hummingbirds have a diet of sugar-rich nectar that is very high in calories to fuel the expense of flight.

◀ Seabirds often have to fly long distances, so in order to save energy they make as much use of the wind as they can.

◀ Larger birds are less agile in general, but the bald eagle (*Haliaeetus leucocephalus*) can fly accurately enough to reach down with its talons to scoop fish from the seas.

bird would have eight times as much muscle and so would be able to produce eight times the power to beat its wings as the small bird, but things are not that rosy. The efficiency of power output from muscles declines with muscle size, so the larger bird will get less than eight times the power out of muscles that are eight times as big. The problem here again relates to surface area. Muscles generate their power from the shortening of muscle fibers, and as muscles get bigger, these fibers get bigger and so can store and release more power. However, the power released by a fiber is proportional not to its total volume, but to its cross-sectional area. As we saw above, surface area increases more slowly with size than does volume, so a muscle that is eight times the volume yields less than eight times the power. This explains why I huff and puff when carrying a package that's only a third of my mass, yet an ant can very comfortably carry ten times its own mass for long distances. The bigger you get, the comparatively weaker you become.

Massive Metabolisms

I weigh 175 lb (80 kg) and I eat a lot! If I was even more eccentric than I already am and had 2,000 pet hamsters, they would weigh the same as me but they would eat a lot more—probably six or seven times as much as I do. This is a general trend: bigger animals need to eat more than smaller ones, but their appetite increases less than proportionately with their mass. This is because metabolism scales less than proportionately with mass.

Your metabolism is the energy that you use up in all your activities—for example, my metabolism right now, while I am sitting and typing is about 200 W, when I am asleep it drops to 100 W, and when I am working very hard at the gym it goes up to perhaps 1,000 W (which is only half the rate of energy used by a hair dryer). If we measured the mass and the metabolism of a range of animals and plotted them on a graph, we would find that the larger animals do have a higher metabolism, but that generally speaking, metabolism changes with mass to the power 0.75. I don't want to get too bogged down in the mathematics of powers, but the crucial thing about the figure 0.75 is that it is positive, meaning that metabolism increases with mass, but it is less than one, meaning it increases slower than proportionately.

This is good news for larger animals, because it means that while they do need to find a lot to eat, they don't need to find quite as much as you might expect. There is more good news, related to the fact that animals store fat to fuel their metabolism when food is scarce. As we scale up an animal in size, its fat store increases proportionately with its mass—for example, taking up 20 percent of its mass regardless of size—but metabolism (the rate at which the fat store is used up) increases slower than proportionately with mass. Thus, larger animals can survive for longer on their reserves. A blue whale (*Balaenoptera musculus*) can feed up in the summer and then live off those reserves all winter, whereas a shrew has to go out and find food every single day, no matter how much snow there is on the ground.

▼ Leafcutter ants are farmers, feeding vegetation to a fungus that they keep in their nest and then nibble on as it grows.

▼ This apple core is much heavier than the ant carrying it, but by working with others, ants can even transport rodents that weigh many times their own weight back to their nest to eat.

▲ Elephants love water as a way to keep cool.

Metabolism and Surface Area

As we have seen, both metabolism and surface area increase with increasing mass, but both do so less than proportionately—and importantly, metabolism does so a little faster than surface area. This has huge ramifications. To illustrate this, bear in mind that the rate at which an animal absorbs oxygen increases with the surface area of its lungs, and the rate at which it absorbs energy from the food it digests scales with the surface area of its digestive system. So, if we scale up the animal but otherwise keep it the same, it is going to struggle to survive, as its metabolism—and hence its need for oxygen and food energy—is going to increase faster than its ability to absorb these essentials for life. This is the reason why the inside of a deer isn't quite as simple as a scaled-up mouse— as animals get bigger, they need to become more efficient at breathing and digesting.

As I sit here typing, using energy at a rate of about 200 W, some of that energy is being put to good use (such as moving my fingers across the keyboard, digesting my lunch, and making new blood cells), but about 90 percent is lost as heat energy. Some of that heat is used to keep my body temperature at about 98°F (37°C), but most of it is just released into the environment—demonstrated by the fact that if I stand up, the cushion I am sitting on would feel warm. From the discussion earlier about surface area and volume, it should be clear that keeping warm is a big challenge for small organisms—but equally, avoiding overheating is a challenge for large ones, because metabolism increases faster than surface area with increasing mass. That's why elephants have large ears that they flap to help them shed heat.

▼ Metabolic rate increases with size but less than proportionately—as approximately mass to the power 0.75— so that if a cow is ten times as heavy as a goat, we would only expect its metabolic rate to be about six times as high.

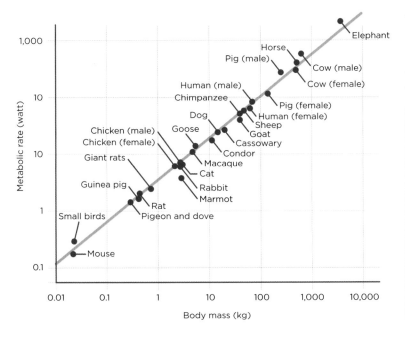

▼ We sweat to cool down when we exercise, but that means we must replace the water we lose.

Let's Get Physical

We need to look at a few more general concepts of physics to help us understand some trends in the natural world that crop up regularly throughout this book. Schools often teach physics, chemistry, and biology separately, but to understand biology fully you really need to know a little bit about the other subjects too. I have managed to avoid chemistry in the issues I discuss in this book, but we do need to focus a bit more on physics.

Why Elephants Don't Run

We discussed earlier in the chapter (see page 14) that a flying bird must counteract the force of gravity through producing upward lift with its wings. The force of gravity acts on all things—for example, as I stand up now, the force of gravity is acting to push me downwards. This force is resisted by my legs, which are strong enough that they don't break, buckle, or bend under the strain of the weight of my body (the force of gravity acting on my mass). Sometimes, however, you see the legs of weightlifters buckling under the tremendous added strain. A weightlifter's legs are generally thicker than mine, which helps them because the ability to withstand pressure increases in relation to cross-sectional area. I'm sure you know where I am heading here: the strain placed on the legs of a bigger animal is greater than that placed on a smaller animal, because mass increases faster with increasing size than does the cross-sectional area of the legs. Thus, elephants keep their legs straight as much as possible to reduce the strain on them, and they don't run for the same reason—they can move fast, but they do so at a walk, always with at least one foot in contact the ground.

▲ A weightlifter not only needs strong arms to lift the bar, but also strong legs to bear all the weight.

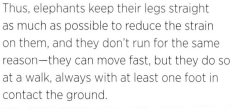

▶ To reduce strain on their long legs, elephants walk quickly rather than run, but when they charge they are surprisingly quick and can cover the ground faster than a person can run to safety.

Why Whales Are So Big

We have discussed several ways in which counteracting the force of gravity is a challenge for larger animals. But there is one way to reduce the effect of gravity hugely—go for a swim. You will have noticed that water supports your body and you can float on the surface, especially if you keep your lungs inflated as much as possible. The force of gravity acting on any object depends on the density of the object compared to the density of the fluid surrounding it. Animals like us are hundreds of times more dense than air, but actually pretty similar in density to water. This is not particularly surprising because a large proportion of our body is water. The upshot of this is that gravitational problems that are an issue on land and in the air are hugely reduced in water. Thus, it isn't altogether surprising that the biggest animal ever—the blue whale—lives in the sea.

Long Legs and Long Levers

Large size generally increases speed of movement. For example, I could run a lot faster than my daughter when she was five because my legs were a lot longer, so I covered a lot more ground with every stride. It is often claimed that efficiency of movement increases with size, such that I would require less energy to walk one mile

than a group of hamsters weighing 175 lb (80 kg) would spend covering the same distance. However, I don't think the evidence for this argument is strong, so although you will find "cost of transport" given as an advantage of large size in some books, you won't find it here.

To experience one last aspect of physics, hold a heavy suitcase (or anything fairly heavy) in both hands but with your arms bent, so that the suitcase is close to your chest. Next, stretch your arms out in front of you so that the suitcase is farther away from you. You should find that when you stretch your arms out, the suitcase feels heavier and it is more difficult to hold it there. By stretching your arms out, you have effectively increased the distance from the fixed pivot point (your shoulder) at which the force of gravity is acting on the suitcase. To a physicist, you have increased the length of the lever arm, thereby increasing the turning moment generated by the force of gravity acting on the heavy weight. The key message here is that holding something heavy far away from your body is hard work. Hence, elephants and rhinos, which have heavy heads, have short necks, and long-necked animals like camels and horses have strong tendons and a fair bit of muscle to support their heads. This concept is important in the next chapter, where we consider the really long necks of some dinosaurs.

▼ A camel's long legs aid travel over great distances, but as a result the animal needs a long neck to reach down to graze and drink.

Some Ecological "Rules"

So far, we have discussed trends in organisms that are driven by how their size influences their interaction with the world around them. Here, we will complement this with a look at a couple of trends that relate to how animals interact with other types of animals.

◀ Rabbits turn only a small fraction of the food they eat into growth and reproduction; most is used to fuel their day-to-day metabolism.

Big, Fierce Animals are Rare

What I really mean here is that (as a generality) bigger animals live at a lower population density than smaller animals, and (as another generality) predators live at a lower population density than their prey. The first of these is easy to understand. Imagine a meadow that's a couple of acres in area, or about the size of a soccer pitch, then imagine how many donkeys you could keep on it—in other words, how many donkeys could have their nutritional needs met by the grass growing in your meadow. Depending on the weather and soil quality, the answer is probably somewhere between two and eight, so let's say five. Now imagine you wanted to keep rabbits instead of donkeys, in which case there is likely enough grass for fifty of them, or you wanted to keep hamsters, in which case your meadow could probably support 500. The reason for this is clear from what we have already seen in this chapter: a larger-bodied animal like a donkey has a higher metabolism than a rabbit or a hamster, so it needs a larger area of grass to support itself. The

same would be true for different-sized fish in a pond or for plants growing on a patch of ground. Bigger organisms space themselves out more—that is, they live at a lower population density. Since the space in which organisms have to live in is finite (say, for hamsters in a meadow or fish in a pond), this tends to mean that the populations of larger-bodied animals are smaller—which is why there are a lot more termites in Africa than there are elephants.

Food Chains are Inefficient

Imagine that our grassy meadow is grazed by 500 hamsters, then consider how many owls this population of hamsters could support. Probably one small owl could just about get by on our hamsters. And this leads us to the second trend: predators are generally much rarer than their prey. The reason for this is that energy is lost from the system as you move up a food chain. Hamsters are inefficient at turning grass into new hamsters, as most of the energy they

▲ Owls generally need a large territory in order to catch enough rodents to fuel their high metabolism.

gather from the grass they eat is used up in their metabolism and vented off as heat, and only a small fraction is invested in growth and reproduction. Owls are similarly inefficient at turning hamsters into new owls, which is why they are at the end of their food chain. If there was a mythical predator that supported itself on owls, it would need to range very widely indeed (over hundreds of acres) and keep that territory all to itself in order to have enough owls to support itself. So, the higher up a food chain an animal sits, the rarer it is.

Since predators often tend to have larger bodies than their prey, both trends often apply at once, and so big predators are really rare indeed. This means that if you go for a walk in the woods at night, you are much more likely to see a rabbit or a deer than a wolf or a bear.

Islands are Odd

Remote islands tend to have fewer species living on them than an equivalent-sized piece of continental land, because it's harder for colonizers to travel across the ocean to set up home there. This means that there is often less competition and less risk of predation for populations of animals on islands, which in turn often causes island populations to evolve to be a different characteristic size from their mainland cousins. Small animals like insects and rodents tend to be surprisingly

large on islands, while large-bodied animals like elephants and rhinos tend to be smaller on islands (if they are there at all). Mainland rodents are small in part so that they can squeeze into burrows to escape when chased by a predator, but if there are few or no predators on an island, then they are released from this pressure and can grow bigger. Similarly large-bodied animals need not grow so big, because they have less need to defend themselves from their predators through sheer size.

▼ The lack of spiders on the Hawaiian Islands left a niche for organisms that catch insects; this has been filled by some caterpillars that are otherwise normally strictly vegetarian.

Chapter 2
DINOSAURS

Whenever anyone thinks of giant animals, there is a very strong chance dinosaurs will spring to mind. This is completely justified. The biggest land animals ever were herbivorous dinosaurs, which are the focus of the first part of this chapter. The biggest predators ever to roam the lands were dinosaurs, too, and we finish the chapter with a good look at these terrifying beasts. Along the way, we also pause to consider why it wasn't just the occasional dinosaur that was massive—huge size was, uniquely, the rule for this ancient group of reptiles.

Sauropods—The Biggest Land Animals Ever

In a book about giants, we should probably start with the biggest land animals that ever lived—the sauropod dinosaurs. They were widely distributed across the world from around 210 million years ago to the end of the dinosaur era 66 million years ago. No matter where they were, the sauropods were generally the biggest beasts in their habitat. Although they all shared a similar body plan—a long neck and tail, and a quadrupedal stance on column-like legs—there were many different sauropods, and they varied in size and, subtly, in lifestyle.

Staggering Sauropod Statistics

It wasn't just that there was one really large sauropod species; they were large as a group, and routinely massive. Some of the commonest sauropods weighed 33–44 tons (30–40 tonnes), and some were likely as heavy as 75–100 tons (70–90 tonnes). As we will see in this chapter, dinosaurs as a group were unusually large, and a few non-sauropods also produced exceptionally large individuals. Some *Triceratops* might have weighed 14 tons (13 tonnes), for example, and some hadrosaurs (duck-billed dinosaurs, like *Magnapaulia laticaudus* and *Shantungosaurus giganteus*) probably also reached that weight or a little more, as did the very largest of the terrestrial mammals (see Chapter 3). However, the sauropods were really in a class of their own in terms of size.

Brachiosaurus altithorax is the most studied species of its genus, and is generally considered to have been 85 ft (26 m) long and weighed about 60 tons (55 tonnes) when fully grown. *Diplodocus carnegii* was a very similar length, but a smaller-bodied beast

with a proportionately very long neck and tail, and so probably weighed "only" 13 tons (12 tonnes). *Brontosaurus excelsus* (which some scientists now consider should be transferred to the genus *Apatosaurus*—but that's a different story) was 72 ft (22 m) long and weighed 28 tons (25 tonnes).

However, the real giants are ones whose names are less well known. The titanosaurs were the last surviving group of sauropods and contained individuals in a number of genera (including *Patagotitan*, *Argentinosaurus*, *Alamosaurus*, and *Puertasaurus*) that are thought to have been 120 ft (37 m) long and weighed about 77 tons (70 tonnes) or more. *Patagotitan mayorum* has a good body of fossil evidence behind it, and although these discoveries are quite recent and have yet to be investigated fully, it feels increasingly safe to assume that adults of this species weighed about 77 tons (70 tonnes). All of these massive dinosaurs have been discovered relatively recently, so hopefully there will be further exciting finds to come.

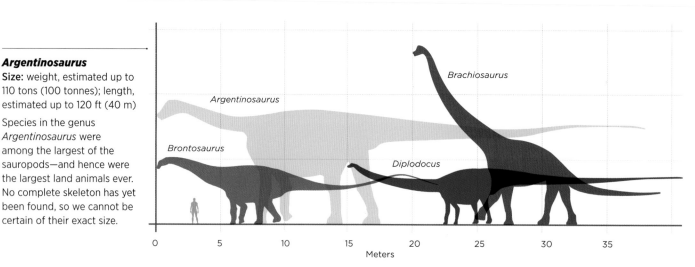

Argentinosaurus
Size: weight, estimated up to 110 tons (100 tonnes); length, estimated up to 120 ft (40 m)

Species in the genus *Argentinosaurus* were among the largest of the sauropods—and hence were the largest land animals ever. No complete skeleton has yet been found, so we cannot be certain of their exact size.

▲ At 85 ft (26 m) in length and 60 tons (55 tonnes) in weight, *Brachiosaurus altithorax* was big, but not giant by sauropod standards. Unlike *Argentinosaurus*, an almost complete skeleton of this species has been found, so we can estimate its size more accurately. It had a particularly long neck for its size (three times as long as a giraffe's, at 30 ft/9 m), which many scientists believe would have allowed it to browse high in the trees.

There are claims of sauropods that are even bigger still, but these have yet to be rigorously explored. One genus to watch out for is *Bruhathkayosaurus*, whose only known species, *B. matleyi*, is estimated to have weighed 80–100 tonnes. You might also hear mention of the 130-ton (120-tonne) *Amphicoelias*, although in this case all the fossil evidence for the type species has been lost. The willingness of scientists to trust in sketches and field notes alone as the basis for claiming it as the largest land animal ever are, understandably, not high.

A Cautionary Tale

You might read reports that *Sauroposeidon* was the tallest sauropod and could stretch its head up 60 ft (18 m) above the ground (about six stories). This may well be true, but the estimate is based on scaling the animal up on the basis of the four neck vertebrae that have been found, assuming it was like similar species for which there are more complete skeletons to work from. However, it is not easy to decide which species are best to use for comparison purely on the basis of just four neck bones, and impossible to know how fair the comparison is (i.e. how similarly shaped the

species were). It is therefore important to remember that many of our ideas about what extinct animals were like are long on (educated) guesswork, and often short on hard evidence.

As an aside, those four neck bones were originally discovered in 1994, but were initially assumed to be a part of a fossilized tree trunk, which gives you an idea of how thick they are. They were considered to be fairly uninteresting and were placed in storage, and their true significance did not become apparent until five years later, when the scientist responsible for them suggested one of his students take a closer look at the "tree trunk."

Benefits of Being Big

There is a lot to be gained from being big. As you become larger, fewer and fewer predators have the physical ability to subdue and kill you, and even those that do may consider that the risks that they themselves are injured are just not worth it. Today, almost no predators try to kill full-grown male African bush elephants (*Loxodonta africana*); only lions (*Panthera leo*) take on the pachyderms, and then only very rarely.

Sauropod Biology

We should take a little time to ponder why the sauropods grew so big. That is, we should consider what the benefits of large size were, what challenges existed that had to be overcome, and what set of circumstances allowed this group to grow bigger than any other. We will find that integral to the functioning of these beasts was a rather tiny head at the end of a very long neck, and consider the controversy about how the sauropods actually used this feature.

▲ Top: Elephants use their height and long trunk to feed from tall trees.

Bottom: Elephants also use their bulk to shake fruit off branches or even uproot whole trees.

Herbivores gain numerous feeding benefits from being big, not least because they can access food that smaller animals cannot. For instance, giraffes can nibble new leaves on trees that are simply too high for any other animal to reach, and elephants can access the tops of trees simply by using their tremendous bulk to knock the tree over. Having a big body also means an animal has lots of room for internal organs, allowing it to digest large quantities of less nutritious food slowly over a long period of time. Finally, when food or water are spatially constrained, large animals can use their size to muscle competing animals out of the way. For example, if a small animal is drinking from a waterhole and an elephant comes along, it would get out of the elephant's way no matter how thirsty it was.

Large size also offers buffering in adverse environments. A camel can walk all day in the hot sun because it takes a long time for its body core to warm to a dangerous level, whereas a tiny rodent in the same environment can risk being out of its burrow for only a few minutes before it overheats. The argument outlined in Chapter 1 about the rate of metabolism increasing slower than proportionately with size means that larger animals can generally last longer without feeding, and this can help them get through contrary weather. A longer stride associated with large size also makes moving over long distances easier and moving to avoid challenging conditions more feasible.

The above benefits mean that larger individuals might be more attractive as a potential mate compared to others of the same species, so there may be reproductive factors selecting for large size in at least one sex. This is especially true in species where males compete physically for access to females. With all these advantages to being big, you have to wonder why there are so many small animals. Clearly, there must be real costs too.

◀ Sauropod footprints on what would have been a beach 66 million years ago but that has since been tilted vertically through geological processes. Note how closely the feet were held together under the body—all dinosaurs held their legs under them, unlike the sprawling gait of other reptiles such as crocodiles.

Costs of Being Big

There are several drawbacks to being large, not least the physiological costs. Sauropods, for example, would have expended a lot of energy and put a lot of strain on their joints just to stand upright. It is unlikely that they were runners, because of the huge pressure this would have put on their feet when they landed. It may well be that they avoided steep gradients and loose or slippery substrates, because they could do themselves considerable damage if their great bulk came crashing to earth. And they obviously needed a huge heart to pump blood around such a vast body. All of this adds up to a massive requirement for food. Something the size of a sauropod needs a huge territory to provide it with sufficient food to meet its needs. One of the most surprising things about sauropods is that preserved tracks of the animals suggest they lived in groups. Such a group would have needed a huge quantity of food to eat, so the foraging advantages of large size listed above must really have paid off for these herbivores.

The other major disadvantage to large size is more an issue for populations than for individuals. The larger individuals are, the longer they have to live before they can reproduce. Also, the larger individuals are, the fewer of them can be supported by a given area of habitat (see page 20). Taken together, this means that there is more of a risk of population extinction for larger-bodied animals. To put it another way, the smaller a population, the more likely it is that random incidents of bad luck can tip it over the edge. And the longer the generation time, the slower the population bounces back from such an incident before the next one strikes. If a particularly harsh winter kills off 90 percent of the rodents in an area, their numbers can bounce back in less than twelve months; for animals the size of sauropods, however, it might take a century for the population to recover, and some other setback could easily occur during that time.

Winning by a Neck

Having a long neck was an integral part of being a sauropod. Animals with long legs need a long neck to feed and drink at ground level (think of horses and camels), but the sauropods took this to extremes. Their really long necks offered two advantages, the most obvious being high browsing—feasting in the treetops on leaves that no other terrestrial herbivore can reach. The other advantage of a long neck is that it allows for a wide feeding envelope at a lower level: a long-necked animal can move its head through a really wide arc and eat lots from a stationary position, thus reducing the cost of feeding by reducing how often it has to move its big body. There is plenty of heated discussion in the scientific literature about which of these advantages is most significant, but the likely answer is that the relative importance of each varied between species—some sauropods may not have been able to raise their neck very high, while others may not have been able to sway their necks out to the side.

One issue faced by animals with a long neck is that they can bear the weight of only a small head at the end of such an extended lever. Consequently, sauropods had tiny heads with simple or no teeth, which meant that

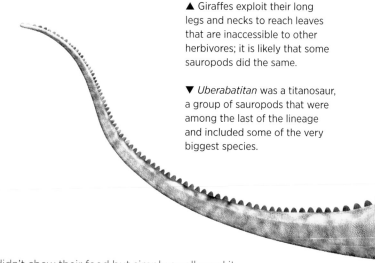

▲ Giraffes exploit their long legs and necks to reach leaves that are inaccessible to other herbivores; it is likely that some sauropods did the same.

▼ *Uberabatitan* was a titanosaur, a group of sauropods that were among the last of the lineage and included some of the very biggest species.

they didn't chew their food but simply swallowed it whole. As discussed above, their big body compensated for this lack of physical processing of food in the mouth by allowing chemical processing over a long period in their huge digestive system.

Dinosaurs are closely related to birds, which as a group have a much more efficient means of breathing than we do (see Chapter 5). It is very likely that sauropods had a similar respiratory system to birds and that this helped them breathe via their long neck, kept their neck lighter by having lots of air in it, and helped keep the animals cool by venting off heat in the air they breathed out.

Super-sized Sauropods

So, why did the sauropods get bigger than any other land animal? As emphasized above, their long neck was pretty key to allowing a giant body size to work for these herbivores, and an avian-like respiratory system was vital in making the long neck work. Larger animals breathe less frequently, but giraffes breathe at a higher rate than expected to compensate for the long distance air has to travel from nose to lungs; the avian respiratory system would have helped overcome the challenge of much longer transport distances in sauropods. In contrast to the massive body size needed by sauropods for chemical digestion (to compensate for their lack of chewing—see above), the other herbivorous dinosaurs had big dental batteries and ground up their food much more thoroughly before swallowing, and so didn't need to be as large.

Compared to mammals, which give birth to live young, sauropods produced eggs—and really small eggs at that. In fact, the eggs laid by giant sauropods were not much bigger than ostrich eggs—about 6 in (15 cm) long and weighing about 3 lb (1.5 kg). The likelihood is that, over her lifetime, a female sauropod could produce hundreds or even thousands of offspring. In comparison, elephants and whales give birth to just one calf at a time and then look after it for several years before producing another one. An elephant that raises more than five offspring in its lifetime has done exceptionally well in reproductive terms. Of course, a large fraction of sauropod eggs and hatchlings perished before reaching the age where they could reproduce themselves, but the sheer numbers produced by a single individual would have given the population a lot more resilience against setbacks than in an equivalent mammalian one.

Finally, there has been a lot of discussion about the metabolic rates of dinosaurs—in particular, whether they had a high metabolism like mammals and birds, or a slower metabolism like extant amphibians and reptiles. The answer is not yet clear, and may be quite complex. Metabolic rate might have varied between different dinosaur types, might have been intermediate between modern birds and reptiles, and might even have changed over the lifespan of an individual. However, there is mounting evidence that at least some dinosaurs had a higher metabolism than would be expected by extrapolating from the reptiles we see around us today. My guess is that at least juvenile sauropods had a high metabolism, and this allowed them to be really active feeders, whatever the weather. In this way, they would have been able to achieve the stunning growth rates that allowed them to grow from a 1 lb (500 g) hatchling to a 33-ton (30-tonne) adult in twenty years or so. A raised metabolism would also help explain why dinosaurs became bigger than any of the reptiles that came before them.

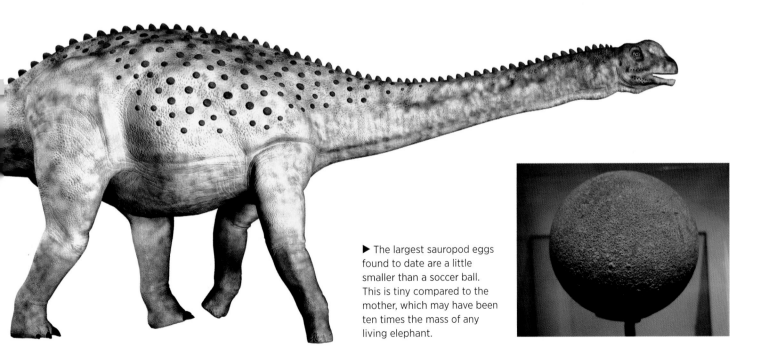

▶ The largest sauropod eggs found to date are a little smaller than a soccer ball. This is tiny compared to the mother, which may have been ten times the mass of any living elephant.

A Gargantuan Group

Among the dinosaurs, it wasn't just sauropods that were big—most species were large-bodied. In contrast, most animals living today are tiny—and I'm not just talking about insects here, but also mammals. As we will see in the next chapter, there are some really huge terrestrial mammals, but these big beasts are highly unusual. Together, the bats and rodents make up more than half the mammal species, and most of them are tiny, weighing less than an ounce (25 g). Here, we look at how dinosaurs differed from animals today in terms of their size distribution across the various known species.

Small is Common

A 2012 study by Eoin O'Gorman and David Hone of Queen Mary University of London in the UK painstakingly collated data on the maximum recorded mass of individuals of each species of vertebrate alive today. For fish, amphibians, reptiles, birds, and mammals, they found that many species had a small body and just a few had a large ones This shouldn't come as much of a surprise. We have just argued that a species with bigger-bodied individuals might be at greater risk of extinction (see page 27). Unlike large animals, smaller-bodied individuals can utilize every nook and cranny of their environment, so there is simply more usable space available to them, and the environment can be divided into more niches among the different co-existing species. So, what O'Gorman and Hone found makes complete sense. When they carried out the same exercise for dinosaurs, however, they found the opposite effect: most species were big-bodied and just a few were small.

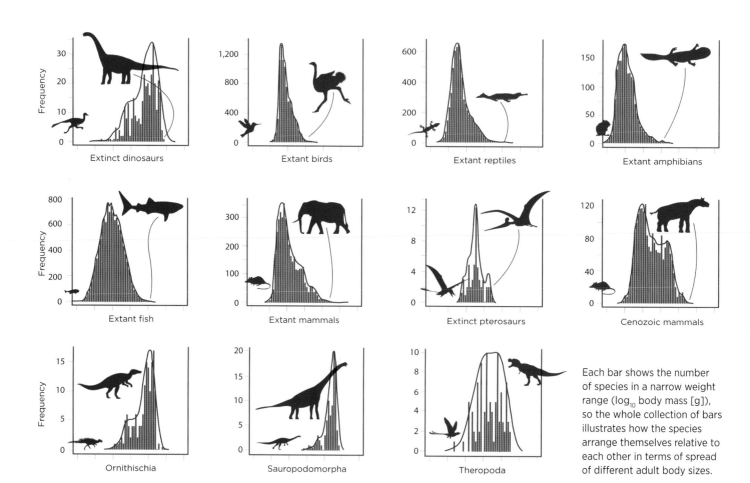

Each bar shows the number of species in a narrow weight range ($\log_{10}$ body mass [g]), so the whole collection of bars illustrates how the species arrange themselves relative to each other in terms of spread of different adult body sizes.

▲ As is the case today, prehistoric waterholes likely attracted many different species of dinosaurs and may have been a dangerous place, where predators waited in ambush.

Fossil Favoritism?

One possible reason for the results is that small-bodied animals do not fossilize as well as large ones, are more readily broken up over time in the ground, or are more easily overlooked by fossil hunters. In other words, the distribution of dinosaurs we are aware of may be a biased sample of the species that actually existed, and the difference in distribution isn't about dinosaur biology, but about the process of finding fossils. To test this argument, O'Gorman and Hone looked at the distribution of two other groups we know only from fossils: pterosaurs (see Chapter 5) and extinct Cenozoic mammals. The data for these showed a distribution intermediate between living species and dinosaurs. This suggests that bias against smaller species is part of the story, but only part—there is something unusual about dinosaurs that led most to attain an enormous size.

The Major Dinosaur Groups

O'Gorman and Hone subdivided the dinosaurs in their study into the three major groups: the sauropods; the ornithischians (e.g. *Triceratops*), which were herbivorous like sauropods and often big—more than a third were heavier than 1 ton (1 tonne); and theropods (e.g. *Tyrannosaurus rex*), most of which were carnivores. They found that the overrepresentation of large-bodied species was stronger in the two herbivorous groups. This led them to theorize why there were few small dinosaurs.

The scientists hypothesized that the herbivorous dinosaurs were selected to be big for all the reasons discussed on pages 25–26. As we saw on page 29, sauropods laid really small eggs, as did dinosaurs in the other two groups, so over their lifetime they changed size dramatically. There would have been lots of small juveniles but few of these would have lived long enough to grow to adulthood. Basically, all the juveniles of big-bodied species took the ecological roles normally taken by small-bodied species. O'Gorman and Hone argued that carnivorous dinosaurs weren't under the same pressure to grow large because most of the animals they feasted on were the small juveniles of the big herbivores, and they didn't need to be big to overpower them.

So, dinosaurs were unusual in that most were really big, not just a few freak species. The dinosaurs did follow the norm, however, in that the biggest individuals weren't predators. Don't despair, though, as there were some predatory dinosaurs that were indeed massive!

Tyrant Lizard

Tyrannosaurus rex (meaning "king tyrant lizard") is probably the most well known dinosaur among the general public, but that's not why it's featured in this book. It gets a mention here because it may also have been the biggest predatory dinosaur—if there were bigger ones, then they were probably pretty similar and closely related. *Tyrannosaurus rex* is definitely the large predatory dinosaur for which there is most fossil evidence, allowing paleontologists to piece together an understanding of what it was like. As we will see, there is considerable uncertainty and disagreement among scientists on aspects of *T. rex's* life, which should give you pause for thought when you consider that speculations are made on other ancient animals on the basis of much less fossil material.

A Giant Named Sue

Paleontologists have found about 50 different specimens of *Tyrannosaurus rex*, several of which are near-complete skeletons. They date back to a relatively narrow time period right at the end of the dinosaur era (68–66 million years ago), and they all come from western North America. Similar, closely related species were around a lot earlier and lived all over the globe, and some of these might have been bigger than *T. rex*, although most were definitely smaller.

Scientists had known about *Tyrannosaurus rex* on the basis of a few bones and teeth for about a century when an amateur paleontologist called Sue Hendrickson discovered the most complete and largest skeleton in South Dakota in 1990. This skeleton was

nicknamed Sue in her honor (although it is uncertain whether the dinosaur was a female or a male) and is on display at the Field Museum of Natural History in Chicago, which bought it at auction for US$7.6 million—the largest sum ever paid for a dinosaur. Museums don't generally have that sort of money lying around, and in this case the Field Museum asked for financial contributions from companies and private donors—the list of those that helped includes Disney and McDonald's. I am glad they gave their support and allowed this amazing find to be displayed to the public and not kept as an ornament in the home of a multimillionaire. I am also thankful that Hendrickson had the skeleton named after her, because she didn't see a cent of the money. The paleontologist was working for a commercial fossil-finding outfit at the

Tyrannosaurus rex
Size: weight, more than 6.6 tons (6 tonnes); length, more than 40 ft (12 m)

This is Sue, among the oldest and largest *T. rex* individuals found to date. How she was found is a complex story with moral and legal implications about who owns fossils.

time, but even that didn't benefit financially—a legal battle that saw the FBI taking possession of the find for a while eventually ruled that the skeleton belonged to the owner of the land on which it was found.

How Big Was Sue?

Because we have about 80 percent of Sue's bones, we have a good idea of how tall and long the dinosaur was—12 ft (3.7 m) tall at the top of the hips and 40 ft (12.3 m) long. One reason why Sue is the biggest fossil *Tyrannosaurus rex* found to date is that when the dinosaur died, it was a lot older than most other individuals discovered to date. If you slice through the bones of a fossil, you can sometimes see annual rings like those in tree trunks. These occur for similar reasons—in temperate regions, the rate of bone deposition changes between summer and winter because the internal temperature of the animal fluctuates, or its diet changes, or its metabolism varies in terms of how it partitions its resources between growth (like bone deposition) and maintenance (like keeping warm). Counting the rings on Sue's big bones reveals that the *T. rex* was twenty-eight when it died. This tells us a number of things. First, these animals were capable of growing really fast; and second, they led pretty short lives for something so big. My guess is that, as soon as an individual showed any sign of weakness (injury, disease, or simply old age), another tyrannosaur would come along, kill it, and eat it!

An Uncertain Weight

Even with a near-complete skeleton there is a lot of guesswork involved in deciding how heavy a dinosaur would have been during its life. It basically comes down to estimating how well developed its muscles were, how much fat reserves it carried, and also what its lifestyle would have been like (and hence how big its different body organs would have been). For Sue, scientists' best guess at the moment is that its mass must have been 9.4–15.4 tons (8.5–14 tonnes). However, this *Tyrannosaurus rex* was likely a fair bit bigger than the average, and a typical adult probably weighed between 5.5–8.8 tons (5–8 tonnes). While Sue was exceptional, this *T. rex* probably wasn't the biggest ever—the skull of another individual measures 60 in (150 cm) long, substantially bigger than Sue's at 55 in (141 cm).

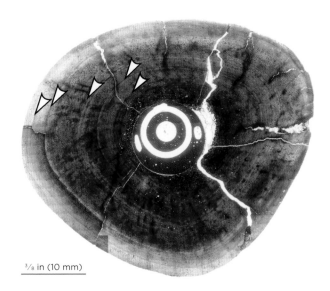

³⁄₈ in (10 mm)

▼ The front limbs of *Tyrannosaurus rex* were comparatively tiny—about the same length as my arms (see page 37).

▲ Just as with tree trunks, we can saw through the large bones of an animal and often find concentric rings, each of which signifies a year of growth. By counting the rings, we can therefore work out how old the animal was when it died or stopped growing.

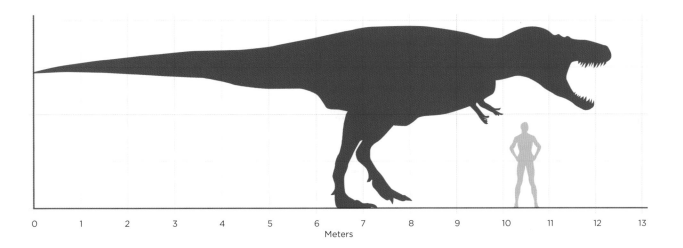

0   1   2   3   4   5   6   7   8   9   10   11   12   13
Meters

Petrifying Predator

Although fossil hunters have found plenty of *Tyrannosaurus rex* bones, piecing together the lifestyle of the animal is far from easy. We have already seen that *T. rex* was a truly massive predator—bigger than any terrestrial mammalian hunter, and likely bigger than any terrestrial predator of any kind. In fact, since *T. rex* became extinct, very few animals of any description that have walked the Earth have been comparable in size.

▲ The primary weapon of a *Tyrannosaurus rex* was its terrifyingly powerful bite, delivered by these large, sturdy jaws.

Strength and Accuracy

The first thing that strikes you when you look at *Tyrannosaurus rex* is its really big head and array of sharp teeth, some of which are more than 12 in (30 cm) long. We can tell from the shape of the skull and its muscle-attachment sites that the reptile would have had a really powerful bite too. However, there is a huge amount of guesswork involved in trying to work out just how powerful this bite might have been, and different scientists have come up with different estimates. From these, it seems likely that a bite force of 6,750 lbf (30,000 N) is a reasonable figure. To put that in perspective, my body weight exerts about 180 lbf (800 N), so being squeezed in the jaws of *T. rex* would feel like having forty grown men standing on your back!

It seems that hiding from *Tyrannosaurus rex* would have been a bit of a challenge too. From inspection of its skull, we can infer that it had a very good sense of smell and exceptional hearing, although its visual acuity seems to have been particularly spectacular. An object big enough that we can just about detect it at a mile (1.6 km) could have been seen from 3.7 miles (6 km) away by *T. rex*. In addition, it had good binocular vision, so it would have had excellent depth perception and been able to strike out accurately with its jaws when it approached its prey.

A Feathered Fiend

For the last twenty years or so, most scientists have agreed that dinosaurs and birds are very closely related. Once in a while, paleontologists find fossil dinosaurs with some of the skin preserved, and in some cases they appear to have had feathers—including some species very closely related to *Tyrannosaurus rex*. So, it would not completely amaze me if *T. rex* had some covering that looked akin to feathers, although my instinct is that the reptiles would not have had anything like the thick plumage we see in birds today.

Some scientists think that dinosaurs had feathers for insulation, in the same way that modern birds do and most mammals have fur. This would imply that they had a high metabolism like mammals and birds. I think it is possible that some dinosaurs did have a high metabolism, but I don't think insulation is the only reason to have a feather-like covering. I think it's possible that display was a key reason behind the evolution of feathers, and that they were important in social signaling, indicating the sex of the animal or its reproductive maturity. To me, this suggests that not only might *Tyrannosaurus rex* have been feathered, but it may also have been quite colorful. However, this remains pure speculation until paleontologists find some well-preserved *T. rex* skin.

▲ Scientists are uncovering increasing evidence that some dinosaurs had feathers. If *Tyrannosaurus rex* displayed this feature, then it likely had only a sparse scattering, since a thick plumage would likely cause overheating in such a giant.

▼ Modern-day terrestrial scavengers follow vultures in flight to find food, so it is possible that *Tyrannosaurus rex* similarly tracked hunting pterosaurs.

Pack Hunters?

It has been suggested that *Tyrannosaurus rex* might have been even more terrifying by hunting in packs like modern lions. I think this is unlikely in adults, yet it doesn't seem too fanciful that small juveniles ganged up to kill animals bigger than themselves. In the case of the adults, there weren't many types of prey a single individual couldn't bring down on its own with its terrifying bite, so there wouldn't have been much advantage in ganging up. Indeed, the cost of doing so would be that the group would have needed a massive territory in order to meet the food needs of all the individuals.

A Cretaceous Conundrum

With an animal as iconic as *Tyrannosaurus rex*, many scientists naturally want to study it and make educated guesses about how it lived. Almost inevitably, the different approaches to challenging questions taken by different scientists have led to some very different conclusions. Here, we touch on a few of these cases, and in each I offer an opinion as to which views I think are the most plausible .

▲ Scientists once imagined that *Tyrannosaurus rex* lumbered along, dragging its tail across the ground. However, the latest studies examining how its bones fitted together suggest a much more athletic lifestyle.

How Fast Were They?

For much of the twentieth century, *Tyrannosaurus rex* was depicted in an upright stance, balancing on a tripod of its large back legs and tail. Scientists imagined that it lumbered around, dragging its tail along the ground behind it. Now, however, we know that its leg joints aren't the right shape for this. Instead, the dinosaur was bipedal (balancing on just two feet, like humans) and held its tail in the air. The tail was huge and heavy, but it needed to be in order to counterbalance the head and body, which were held forward of the legs. Without the big tail, the animal would have continually toppled forward.

The difference between walking and running is that when you walk you always have at least one foot on the ground, whereas when you run there are moments when you are completely airborne. When they charge, elephants actually walk really fast—they don't run as they never have all four feet off the ground. My instinct is that the higher pressure exerted on the legs and feet when running rather than walking would have been too extreme for a beast weighing as much as *Tyrannosaurus rex*, but with its 10 ft-long (3 m) legs it would have been able cover ground faster than most people can run—most estimates of its pace come in at 25 mph (40 kph), equivalent to an elite sprinter

▲ A reconstruction of a more athletic *Tyrannosaurus rex* suggests that its powerful hind legs, with their giant claws, could have been a useful weapon—as were its terrifying jaws.

Predator or Scavenger?

I can't fathom why scientists continually debate whether *Tyrannosaurus rex* was predominantly, or even entirely, a scavenger. It clearly had the equipment to be an effective predator, but as in all large modern-day predators, it would have scavenged from corpses when it could, and it would have tried to steal kills off smaller predators.

Modern vultures are highly unusual in that they are often obligate scavengers. However, the reason for this is that they have evolved a body plan that gives them huge wings for effective soaring, allowing them to fly for long periods in search of a meal at very limited cost (see Chapter 5). This makes them good scavengers, but at a price. Their large wings make them really inflexible, which means they can't be predators—they simply can't fly or land accurately enough to strike potential prey. In animals with legs, no such trade-off occurs between movement for scavenging and movement for predation that there is no reason why any—including tyrannosaurs—would become a full-time scavenger. It seems that *Tyrannosaurus rex* was every bit the ferocious predator of your nightmares!

running in ideal conditions. However, in a mythical world where *T. rex* hunted humans, we might have been saved by our maneuverability. A tyrannosaur was a bit like a supertanker when it came to turning—in other words, it may have been fast in a straight line, but prey could possibly have avoided its bite by continually dodging out of its path. This strongly suggests to me that *T. rex* was not a generalized predator like modern bears, for example, but instead focused on really big prey that could not outmaneuver it.

Why the Tiny Arms?

Although *Tyrannosaurus rex* weighed at least 5.5 tons (5 tonnes), its forelimbs were just 40 in (1 m) long (for comparison, mine are about 30 in/80 cm, long). However, these arms had thick bones, lots of muscle, and a limited range of movement—just the qualities needed to hold down prey while it maneuvered its head to deliver a killing bite. Other theories on why the arms were short are that they helped the dinosaur get up after lying on the ground, and helped it to hold onto a partner during copulation. The limbs may well have been used in these contexts, but to me they look overengineered for anything other than manipulating prey struggling for its life.

▼ It is unlikely that an adult *Tyrannosaurus rex* was nimble enough to catch fast, agile small-bodied prey. Instead, it may have focused on occasionally attacking large prey like these *Triceratops*.

Chapter 3

MASSIVE MAMMALS

The biggest land animals alive today are mammals, and that has been true for the last 66 million years, since the last of the giant dinosaurs died out. In this chapter we look at the biggest of these land mammals, past and present. Of course, the largest animal ever to have lived is the blue whale (*Balaenoptera musculus*), which is a mammal, too. This species is covered in the next chapter, "Giants of the Deep," but to get you attuned to the aquatic mammals, we mention here an often overlooked group that is closely related to elephants. We start, though, with the biggest land animal alive today: the African bush elephant.

Enormous Elephants

When I was a boy and a teacher asked for the name of the largest species roaming the Earth, then the answer "African elephant" would have got you full marks. This is not strictly true, however, as genetic studies have since shown that what we used to think of as the African elephant is actually two species: the African bush elephant (*Loxodonta africana*) and the African forest elephant (*L. cyclotis*). These close relatives separated from one another between 2 million and 7 million years ago, before humans separated from the chimpanzees. Of the two, the bush elephant is slightly the larger, so that will be our focus here.

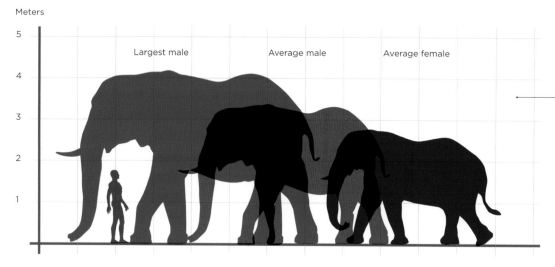

Meters

5

4

Largest male   Average male   Average female

3

2

1

African bush elephant
Loxodonta africana
Weight: up to 11 tons
(10 tonnes)

The silhouettes show average adult sizes and the size of the largest known individual of what is the largest species of land animal alive on the planet.

Elephant Biology

The very largest African bush elephants are almost 13 ft (4 m) tall at the shoulder and weigh a little over 11 tons (10 tonnes). In comparison, the largest African forest elephants weigh 6.6 tons (6 tonnes) and the largest Asian elephants (*Elephas maximus*) around 7.7 tons (7 tonnes). The largest rhinos and hippos (*Hippopotamus amphibius*) come in under 4 tons (3.5 tonnes), so elephants are comfortably the largest land mammals, and the largest of them all is the African bush elephant.

One really unusual anatomical feature of elephants is their large head, which can make up 25 percent of their body weight. Humans also have a proportionately large head, but it makes up only about 7 percent of an adult's body weight. Elephants need a big head to hold the massive dental batteries that are one of the keys to their lifestyle—they can digest even the most unappetizingly tough vegetable matter, such as whole

tree branches or the bark from the trunk. They go through six sets of teeth in their seventy-year lifetime, such is the volume of tough material they have to grind their way through to meet their energy needs. The head is also heavy because it supports the ivory tusks, which are used for display and aggression, as well as for digging for food and water. The large size of the head also makes it handy for pushing down trees, either to allow the animal to access food or simply because the tree happened to be in the way. However, the problem with having such a massive head is that it must be supported by a really short neck; this places the head almost directly above the front legs, thereby allowing the weight to be carried more efficiently. A short neck presents a challenge for feeding and drinking, but evolution has elegantly solved this with the trunk. This allows elephants to eat and drink without having to move their head—the flexible yet strong trunk brings food and water directly to their mouth.

▲ Elephants are strongly social, with females often spending all their life in the same tight-knit family group.

Being large endotherms in a warm climate, elephants face a challenge in terms of overheating. They are surprisingly good swimmers and love to bathe to cool off, and they also use their trunk to spray cooling water or even soil onto their back, but their key adaptation for keeping cool is their big ears. These are well supplied with blood vessels, which combined with their large surface area allows heat to be lost through convection—providing the surrounding air temperature is cooler than the body temperature. Even better, the ears can be flapped, which increases airflow past them but also past the upper part of the body, further aiding heat loss.

Whenever I see an elephant skeleton in a museum, I see it as a humbling lesson about what we think we know about extinct organisms. So much of this is inferred from their bones, because these are much more likely to be preserved than any other tissue. But there are no bones at all in an elephant's trunk or external ears. If elephants were extinct and all we had was their bones, then it would be a much more gifted scientist than me who would have come up with the idea of the ears and especially the trunk. The cautionary lesson here is that there might be some extinct animals that we really don't understand as well as we think we do, because important parts of their anatomy had no bones and therefore have left no fossil trace.

▲ Elephants require huge dental batteries to grind up the enormous quantities of forage they need to keep them going.

▼ If elephants were extinct and all scientists had to work from were their fossilized bones, would they have guessed that the animals had a trunk and huge ears?

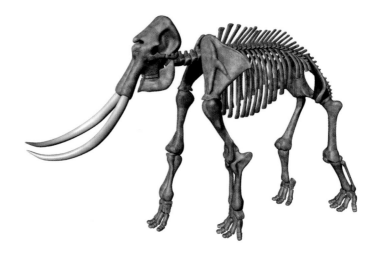

▲ Elephants have very slow reproductive rates but compensate by having extensive, prolonged parental care—a little like humans.

Under Threat

Adult African bush elephants have no natural predators, except perhaps lions (*Panthera leo*) in highly unusual circumstances. The young aren't particularly vulnerable either, given that they spend their time not just with their mother but in a highly cohesive and protective social group of juveniles and older females—most lions would wisely conclude that there are easier ways to find a meal. Even if a

pride of lions did manage to kill an elephant calf, the likelihood that they could consume it without being stamped on by the enraged herd is minimal. Not surprisingly, the main threat to elephants comes from humans—the African bush elephant might be the most common elephant species, but some estimates put the population decline at 8 percent a year, with the very real possibility that it will become extinct within our lifetime. One part of the challenge is its slow reproduction rate. Elephants have a gestation period of twenty-two months, so a female might give birth to a single calf only every five years, since the calf will often still be feeding from its mother up until the age of five (despite increasingly eating solid food for itself from six months onwards). A female might be reproductive from the age of ten, but reaches peak fertility between the ages of twenty-five and forty-five. For such slow breeders, any increase in levels of predation can be serious; sadly, the adverse effects of humans are manyfold.

Anyone guessing that poaching elephants for ivory has a significant negative impact on their numbers would be absolutely correct. Many African countries try hard to curtail poaching, but the challenges are considerable. First, the bush elephant population is widely dispersed and individuals roam very widely, so it is not practical to guard them fully. Second, much of Africa does not have the infrastructure to implement anti-poaching regulations effectively. And third, simple

◄ An elephant herd can include three different generations and a broad diversity of ages.

◀ Ivory poaching remains a serious problem for the future of elephants across large parts of Africa, but so too is the conflict between the animals and farmers who are trying to grow crops without having them trampled and consumed.

▼ There is growing international support for measures to curb ivory poaching.

economics is a barrier to conservation, as there are many people in Africa for whom the possible wealth offered by the extraordinary value of ivory is worth the risk of perhaps even being shot while in the act of poaching. I believe that while demand remains as high as it does, there is simply no way to stop ivory poaching in Africa. What is required is attitudinal change to the desirability of ivory products globally, but especially in the parts of Asia where most of the ivory ends up. There is no ivory in my house; I hope there is none in yours.

Ivory is not the only reason why elephants are deliberately killed. Big game hunting remains a lucrative enterprise across parts of Africa, and lions and elephants are inevitably the targets most sought after by trophy hunters—some people will pay tens of thousands of dollars for the opportunity to kill one of these magnificent creatures. Again, I think an attitudinal change is needed. If you have an elephant's head on your wall, don't bother inviting me to dinner. And don't tell me about all the wealth your trophy hunting injects into Africa—if you care about helping Africans, then give your dollars to a charity that delivers clean drinking water, not to someone who will find you a beautiful animal to kill.

The human population in many parts of Africa is growing rapidly, and that inevitably means more and more land is being taken up by agriculture, by livestock, and by human habitation. As a result, there is more of Africa where elephants are simply unwelcome, which itself produces a number of challenges. For a start, elephants are not great respecters of "keep out" signs and fences, and so conflict with farmers is likely. Investment in affordable and effective boundaries is needed. This is relatively straightforward, as even a low boundary wall is enough to pen them in—you might have noticed on visits to zoos that elephants aren't especially acrobatic. More creatively, beekeeping might be a cost-effective way to keep elephants out of crops, as they are not great fans of the insects. If we can curtail the deliberate killing of elephants and keep them out of areas where they are unwelcome, then there is enough suitable habitat in Africa to support a reduced but flourishing population of the largest land animal on our planet for future generations to enjoy.

Extinct Elephants

Ancestors of modern-day elephants first appeared around 50 million years ago, and for various reasons we know a lot about them. We know more about relatively recent species, because the ground containing their bones is less likely to have undergone catastrophic change, and we know more about larger species, because their bones are less likely to have been damaged and more likely to have been spotted by paleontologists. But the main reason we know a lot about elephant ancestors in particular is because teeth preserve especially well, and these animals (like the elephants of today) had huge dental batteries. Further, their teeth were under strong selection pressure—being vital to survival, and shaped by the longevity, diet, and size of the animals—so tooth shape seems to be a reliable diagnostic of individual species.

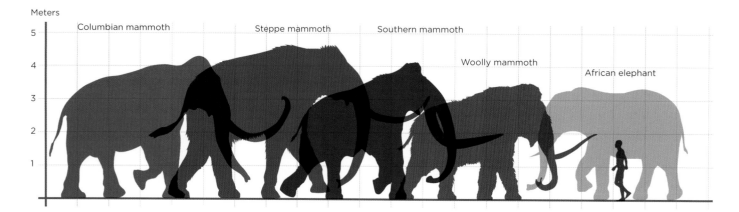

Mammoth Beasts

The most famous extinct relatives of the elephants are the mammoths, which first appeared around 5 million years ago and died out as recently as 4,000 years ago. Not all of them were huge, but the largest—the southern mammoth (*Mammuthus meridionalis*), steppe mammoth (*M. trogontherii*), and Columbian mammoth (*M. columbi*)—were a bit bigger than the largest African bush elephant (see page 40), weighing perhaps 13 tons (12 tonnes). They had a wide distribution across Africa, Asia, Europe, and the Americas. Those from the north had long, shaggy coats (although species from warmer regions did not), and most had proportionately larger and more curved tusks than modern elephants.

We know a huge amount about the more recent mammoths, including the woolly mammoth (*Mammuthus primigenius*), for two main reasons. First, a number of specimens have been perfectly preserved in the permanently frozen substrate in some northern regions such as Alaska and Siberia. When

these are thawed out, we can investigate them as if they died only yesterday, including their soft tissue, their stomach contents, and their DNA—an absolute treasure trove for scientists. Second, these animals lived alongside humans, who depicted them on cave walls and carved their images into portable artifacts. One of the earliest depictions of an animal carved by a human is of a mammoth, fittingly (or ironically) carved from a piece of ivory. Long story short, these animals appear to have been very similar to modern elephants in much of their anatomy and lifestyle.

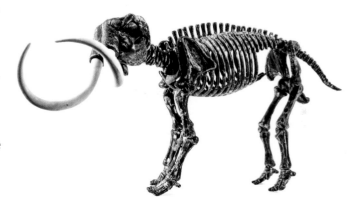

◀ The exceptional preservation of the soft tissue of mammoth remains in permafrost allows us to be certain that the animals were hairy.

Damning Evidence

Why did the mammoths die out? The possible answers come down to climate change alone, humans alone, or a combination of the two. While there have been ice ages and other climate changes over the last 25,000 years as the mammoths (and many other large animals) faded to extinction, these events have been no more dramatic than similar episodes over the previous millions of years, so I don't think we can pin the demise of the mammoths on climate alone. In contrast, humans and our immediate ancestors have changed dramatically—we have become more numerous, have spread more widely across the planet, and, most importantly, have developed technologically. I think the mammoths went extinct because humans became more efficient at killing more of them across more of their range. But this is just my opinion—it isn't fact. There does seem to be a correlation between when humans occupied different regions of the world and when mammoths went extinct there, but this correlation isn't perfect—we can't date either type of event all that accurately in a given location. Cave paintings and other artforms do depict mammoth hunts and some mammoth remains have been found with spearheads embedded in them, but hunting need not necessarily equate with extermination. There is no conclusive evidence that humans had an important hand in the demise of many large animals (including mammoths) in the last 25,000 years, but I would say that the weight of evidence is damning.

Re-creating the Mammoth

Scientists have extracted intact DNA from mammoths and have mapped their full genome, so it is theoretically possible that they could re-create a living mammoth. The genome is 99 percent the same as that of the Asian elephant, and scientists have been successful in adding short strands of mammoth DNA into elephant cells. The pace of scientific change is extraordinary—a century ago we had never even heard of nuclear energy, or imagined heart transplants or sending a man to the Moon—but whether it is possible to re-create a mammoth or not, it will certainly be a challenging and expensive business. If, in a hundred years' time, we have re-created a herd of mammoths in a theme park somewhere but have allowed African bush elephants to go extinct in the interim, then I'm not sure that would feel like a fair exchange. That said, I accept it's an exciting area of biology, and whether or not we should be funneling resources into resurrecting mammoths is at least a debate worth having.

◀ The skeleton of this mammoth emphasizes how similar they were to modern-day elephants—aside from the extraordinary tusks.

▶ One consequence of global warming will be the uncovering of more mammoth remains as permafrost thaws.

North and South American Giants

The mastodons (a group of species collected into the genus *Mammut*) inhabited North and Central America until about 10,000 years ago. They are distant relatives of the elephants and mammoths, having separated from them around 27 million years ago, but they seem to have been very similar to them in anatomy and lifestyle. While mammoths had teeth like modern elephants and were particularly suited to grazing on the plains, mastodons had teeth more selected for tackling woodland foliage. The first examples were discovered when giant teeth were uncovered in various locations in North America in the eighteenth century. Allegedly, African slaves first suggested a similarity between these teeth and those they had seen in elephant remains. Mastodons had a build quite similar to that of the modern Asian elephant, but the largest individuals might have reached 12 tons (11 tonnes). It won't surprise you to hear that I link their disappearance to the sweeping spread of humans across the Americas following their arrival from north Asia around 17,000 years ago.

The gomphotheres were another extinct group of elephant-like animals, which were widespread across North America from 12 million to 2 million years ago. Unlike the mastodons, they did venture into South America when sea-levels dropped about 5 million years ago and a land bridge linked North and South America. What happened then is called the "great American interchange," which saw animals from the south move north and vice versa. The gomphotheres moved south just as they were being replaced by other elephant-like animals in North America, and thrived there until around 9,000 to 6,000 years ago. I won't even bother to tell you why I think they died out.

▼ *Amebelodon fricki* was a herbivorous mammal that inhabited the plains of North America during the Late Miocene (11–5 million years ago). It had two sets of tusks, the lower, flattened pair of which protruded from the front of the lower jaw and is thought to have been used as a shovel to excavate plant tubers from relatively soft ground.

▶ Models of giant mammoths emphasize their rather pointed heads compared to those of extant elephants.

Asian straight-tusked elephant
Palaeoloxodon namadicus
Weight: 11.5 tons (10.4 tonnes)

Recent work suggests this might have been the largest land mammal species ever to have lived.

The Biggest Ancestral Elephant

An important study called "Shoulder Height, Body Mass and Shape of Proboscideans," by Asier Larramendi, was published in 2016; it can be accessed online in full, free of charge, simply by typing the details into your favorite search engine. Proboscideans is the collective name scientists use for the elephants and their ancestors, and Larramendi's study is a comprehensive review of different approaches that have been taken to estimate the size of extinct creatures of this type. Scientists generally have only an incomplete skeleton or sometimes even just a single bone to work from when estimating an extinct animal's size, so they have to make numerous assumptions and then take a guess based on the evidence. Larramendi's paper makes a number of important claims.

The first is bad news for modern-day elephants. Although the largest African bush elephants are generally estimated to weigh around 11 tons (10 tonnes), none of the very biggest individuals that have been shot have actually been weighed. This is understandable, since the equipment required to weigh something this large is not widely available. Instead, measurements like foot circumference are often taken, and from this the weight is estimated. Using these data, Larramendi suggests that weights like 11.5 tons (10.4 tonnes) are actually more likely for the African bush elephant than 13 tons (12 tonnes). That said, this figure still makes the species the biggest land animal currently on the planet.

Larramendi then goes on to present evidence that quite a few proboscidean species were a little bigger than current bush elephants, with weights of around 12 tons (11 tonnes), and that some were quite a lot bigger. For example, he estimates that one individual of the mastodon species *Mammut borsoni* weighed 15 tons (14 tonnes). This animal is thought to have been about 30 years old, and Larramendi suggests it might have grown to as much as 20 tons (18 tonnes) had it lived to a ripe old age.

It seems likely that the largest proboscidean of all was the Asian straight-tusked elephant (*Palaeoloxodon namadicus*), which could be found across Asia from India to Japan until about 24,000 years ago. One partial skeleton found in India in 1905 was particularly enormous, and Larramendi estimates that this animal might have weighed 24 tons (22 tonnes) when alive. This is big news for the proboscideans, as scientists previously generally considered that the largest mammal ever to walk the planet was a 19-ton (17-tonne) ancestor of the rhinos (see page 50).

Remarkable Rhinos

The other really large living land mammals are the rhinos and hippos. Here, we look at extinct rhino relatives as well as the biggest rhinos found today. The ancestors of rhinos were more widely distributed across the globe than their descendants, and included some real giants that were quite different from the rhinos of today.

The Living Rhinos

There are five species of extant rhino and most sources say that the white rhinoceros (*Ceratotherium simum*) is the largest of these. I am not going to dispute this, but there certainly isn't a huge difference in size between the average individual of that species and the average Indian rhino (*Rhinoceros unicornis*), and few accurate records of either species appear to have been kept. Males are heavier than females in all species, and a typical male white rhino might weigh 2.5 tons (2.3 tonnes), compared to 2.4 tons (2.3. tonnes) for the Indian rhino. It's pretty easy to imagine particularly large individuals of either species reaching 4 tons (3.6 tonnes), say, but even weights of 4.4 tons (4 tonnes) or above seem plausible. As it has the largest recorded average size, we will focus on the white rhino. But before we leave the Indian species, I have to confess that I think it looks the coolest—it has thick skin folds that, to me at least, make it look heavily armored.

The White Rhino

Something odd happened with the naming of the two African rhinos—the white rhino and the black rhino (*Diceros bicornis*)—because color is not in any way a reliable trait to distinguish them. The two animals are pretty similar, being very closely related and having diverged from one another sufficiently recently that they can still interbreed. The white rhino is generally bigger, but you might still struggle to differentiate between an adolescent white and a full-grown black. The best way to tell them apart is by mouth shape: the white is a grazer, with a wide lower jaw and mouth, while the black browses on trees and shrubs, and has a small mouth for carefully stripping branches.

There are two subspecies of the white rhino, and here the naming is more accurate, as the northern subspecies (*Ceratotherium simum cottoni*) was spread across eastern and central Africa just south of the

Indian rhinoceros
Rhinoceros unicornis
Weight: up to 4.4 tons
(4 tonnes)

While Indian rhinos are not quite as large on average as African white rhinos (*Ceratotherium simum*), I find their armor-like folds of skin visually striking.

▲ Individual rhinos can often be reliably distinguished by the size and shape of their horns.

Sahara, and the southern subspecies (*C. s. simum*) is mostly confined to South Africa. The southern white rhino's historical range also included other southern African states, and it has recently and successfully been reintroduced to a number of these. You might have noticed a change in tense in my description of ranges; this was deliberate, because the northern subspecies may well be extinct in the wild and only two captive individuals exist, both females and both in Kenya. There have been attempts to breed these two females with a male from the southern subspecies, but these have been unsuccessful, and it appears that the females are infertile. This means that the subspecies

might be functionally extinct—although we haven't quite given up hope altogether. No individuals have been reliably reported in the wild since 2008, but these animals can live to be fifty, and although they are pretty huge, the last place where they were seen is in the 2,000-square-mile (5,200 km^2) Garamba National Park in the Democratic Republic of Congo. It is just about possible to imagine how a couple of rhino could have grazed quietly for the last ten years in a remote corner without being noticed. To end on a happier note, the southern subspecies is now thriving after nearly dying out a century ago; today, the population numbers about 20,000.

White rhinoceros
Ceratotherium simum
Weight: up to 4 tons (3.6 tonnes)

On average, adult white rhinos are a little bigger than adult Indian rhinos.

Black rhinoceros
Diceros bicornis
Weight: up to 3.2 tons (2.9 tonnes)

In this species the front of the face and mouth are characteristically narrower than in the white rhino.

Extinct Giant Rhinos

Rhinos have been around for tens of millions of years and were much more widely distributed across the globe than they are today. Our interest here centers on the biggest of these, members of the genus *Paraceratherium*, which ranged widely across Asia and into Europe between 34 million and 23 million years ago. They were bigger than modern rhinos and even modern elephants, as we shall see.

It seems that these rhino ancestors browsed on soft leaves, and from their skull morphology it looks like they had a super-muscular upper lip or even a proto-trunk like a modern tapir to help with this. Unlike modern rhinos, they had a relatively long neck and this, combined with longer elephant-like legs, would have allowed them to reach up high in their quest for food. Having such a long neck meant that these animals couldn't have a massive head (see page 19), and they didn't have quite such huge teeth as modern elephants, suggesting that they weren't interested in grinding up very fibrous material but instead focused on soft leaves. They were also hornless, which would also have saved weight from the head.

Plenty of popular science books describe *Paraceratherium* species as the biggest land mammals ever to have walked the planet. As we discussed earlier (see page 47), Asier Larramendi re-evaluated the possible weights of large elephant ancestors, suggesting that some of them exceeded 22 tons (20 tonnes). The earliest estimates, from about a century ago, put the mass of *Paraceratherium* individuals anywhere between 22 tons (20 tonnes) and 33 tons (30 tonnes), but since then these estimates have been revised downward. Scientists are slow to criticize others in print directly, but in 1993 Mikael Fortelius of the Finnish Museum of Natural History and John Kappelman of the University of Texas produced the most comprehensive effort to estimate the size of these giant rhinos by different means. The last word of the title of their paper, "The Largest Land Mammal Ever Imagined," can be seen as the politest way possible of suggesting that the previous calculations were overinflated. They estimated the mass of individuals in 92 different ways (depending on what bones and assumptions were used) and hardly ever ended up with a figure above 16.5 tons (15 tonnes). Larramendi focused on what he considered the most reliable sets of assumptions and came up with a maximum weight for *Paraceratherium* individuals of around 19 tons (17 tonnes).

From this, we can conclude that *Paraceratherium* species were certainly among the very largest land mammals ever, but were probably not the very biggest of all, given that there were likely some proboscideans that topped 22 tons (20 tonnes). It's not really clear why these giant rhinos died out, but there seems to have been expansion of proboscideans into Asia around this time, and my guess is that these outcompeted the giant rhinos, perhaps because of their ability to eat fibrous vegetation as well as the softest leaves. In addition, overheating could have been more of a problem for the larger *Paraceratherium* species, as they lacked the big ears of elephants. Warming of the climate, combined with competition from proboscideans that had traits protecting them against overheating, might well have done for the giant rhinos.

Meters

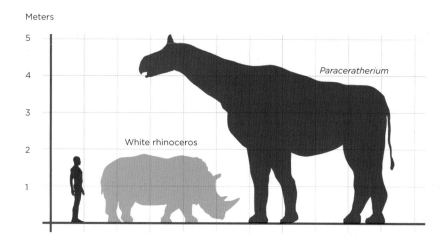

White rhinoceros

Paraceratherium

Herbivore Heavyweights

We discussed in Chapter 1 why we would expect
herbivores to be bigger than carnivores, but why be
a giant herbivore like a rhino rather than a small one
like a rabbit? The answer to this is not as clear as
scientists once thought. It has been assumed that large
size allows greater digestive efficiency, because big
animals can hold more food in their system for longer.
But careful study by Marcus Clauss of the University
of Zurich and others has demonstrated that this
does not to appear be the case when we look at the
very largest herbivores living today, so the selection
pressures that caused very large size to evolve must
be found elsewhere. In terms of foraging, it is likely
that larger animals can, on average, last longer on their
body reserves than smaller animals (since metabolism
increases less than proportionately with mass; see
page 16). Large size might also make long-distance
movement in search of better food more practical, and
it allows preferential access to any food source through
the intimidation of smaller-bodied animals and to high
browse through reaching up or knocking trees down.
Finally, it might be that selection for reduced predation
risk drives some species to evolve huge sizes such that
nothing much bothers them.

We should also ponder why the largest herbivorous
dinosaurs were bigger than the largest mammalian
equivalents. One possibility is overheating, such that a
mammalian metabolism cannot be sustained at the size
of the largest sauropod dinosaurs without "cooking" the
animal concerned. However, I suspect the real reason
relates to their very different reproductive strategies.
It seems inevitable in mammals that the larger they
get, the slower the rate of offspring produced, whereas
sauropod dinosaurs produced many relatively small
eggs at any one time. Admittedly, this would mean
that sauropod youngsters would have suffered high
mortality (say, through predation), but if there were
repeated environmental shocks that wiped out a chunk
of the population, then the sauropods were better able
to bounce back quickly (especially if the environmental
shocks also took out many of the predators) than large
mammals. What might ultimately set the upper limits
to mammal size is how quickly a population can recover
from a setback such as a drought or flood before
another environmental shock occurs. The baleen
whales seem to contradict this, but they rely on a
bonanza of food that is much more reliable year on
year and is so overabundant that competition between
feeders is irrelevant.

Huge Hippos

We round off our discussion of megaherbivores with the hippo, as an example of a different solution to the problem of overheating at large size. Hippos spend the daylight hours submerged in water, coming out only during the cool of the night to graze on nearby grasslands.

Hippopotamus
Hippopotamus amphibius
Weight: up to 3.5 tons (3.2 tonnes)

Although it is a herbivore, the hippo attacks and kills humans—males can be fiercely territorial of their stretch of river, and getting between a female and her offspring is definitely not recommended.

Hippo Biology

The hippopotamus is most closely related to the whales and dolphins, but split from them 55 million years ago. It is widespread is sub-Saharan Africa wherever there is shallow water and gently sloping banks near grassland. Ancestors of modern-day hippos ranged across Europe and Asia, as well as Africa, and although some were a little bigger than individuals found today, they were not hugely so. For this reason, we focus here just on the living species.

Male hippos are larger than females and apparently all hippos grow continuously throughout their life. Large individuals weigh around 4,400 lb (2,000 kg), but there are plausible reports of exceptional individuals reaching between 6,000 lb (2,700 kg) and 7,000 lb (3,200 kg). Strangely for an aquatic animal, hippos are not great swimmers. They are negatively buoyant and stand on the bottom, rarely straying into waters so deep that they have to swim. They love a depth where they can touch the river or lake bed with just the top of their head showing, since their ears, eyes, and nostrils are all high on their head. Despite not being great swimmers, they do everything in the water except eat—they mate in the water, give birth in the water, and suckle their young during the day in the water. They also sleep in the water and, amazingly, can do so in water that is too deep for them to keep their nostrils permanently above the surface. In these circumstances, the sleeping individual will push itself upward with its forelegs every few minutes so that its head bobs above the water. The hippo's nostrils then open, it exhales and inhales, and it submerges again—all without waking.

▲ Hippos yawn as a form of social signaling, mostly indicating their dominance.

Hippos in South America

Apparently, during their 55 million years of existence hippos have never spread into the Americas. However, this has changed in my lifetime thanks to recreational drugs. The notorious Colombian drug lord Pablo Escobar had four hippos in his private zoo. After his death, they were deemed too difficult to catch and rehome, so they have been allowed to roam free in his now untended estate—and the four have become at least forty and started to spread their geographic range. A strategy will be required to check the population or to make sure the local people are comfortable with these new residents. No matter how dangerous hippos are in reality, it will be a long time before this population kills anything like as many people as the cocaine that funded their strange introduction to the Americas.

▼ Hippos can walk along the bottom of a waterbody completely submerged for several minutes.

Are Hippos Dangerous?

It is often stated that hippos kill more people in Africa than lions. I don't think this is easy to prove, since the fate of people who go missing often remains unknown. However, there are more than 100,000 hippos compared to perhaps 20,000 lions, and hippos definitely do attack boats and people. They do this for two reasons. First, males are highly territorial in the water—they fiercely defend a stretch in which they maintain priority or exclusive mating rights. This makes them highly aggressive toward anything that passes by, including boats and swimmers. Second, although adult hippos are generally avoided by even the biggest lions and crocodiles, the young are vulnerable if left alone. All adult hippos, and especially mothers, are very protective toward possible threats to youngsters, so if you happen to blunder into a group of feeding hippos at night and get between a mother and her offspring, then bad things are quite likely to occur.

The sheer bulk of a hippo is enough to flatten and kill you, but they also have really large front teeth that they like to show off with a gaping yawn. The teeth aren't used for feeding at all, but for fighting, especially in contests between males. The animals can have quite a violent squabble without doing too much damage to one another, because their skin is 2 in (5 cm) thick. This

protective skin and the formidable teeth are further reasons why predators prefer to find easier meals than hippos. However, the teeth can also lead to their demise at the hands of humans, as they have substantial value as ivory. Hippos have also been killed for food, but the two biggest threats they face are being killed to protect crops that they might otherwise graze on, and habitat loss to an expanding human population (especially given their need for both swallow water with gently sloping banks and good-quality forage nearby).

Extinct South American Giants

▶ The tough shell of the glyptodons must have been heavy—as indicated by stout, weight-bearing legs. They also had a bony head and tail, similarly reinforced to withstand attack.

In times past, South America was home to a couple of unusual giant herbivores. One of these was like an armored car, while the other was one of the largest herbivores ever known—and both still have much smaller descendants living today.

▲ Modern-day armadillos are ancestors of the giant glyptodons, but are much smaller and have a leathery exterior rather than a bony shell.

The Armored Glyptodons

Glyptodons lived in South America from around 2.5 million years ago until around 10,000 years ago. They were close relatives of modern-day armadillos, but took the protective armor of these animals to a whole new level and also reached massive proportions. For some reason, they are usually described as being about the size of a Volkswagen Beetle—I think the domed shape of their shell is partly responsible for the analogy popping into people's heads. There is no doubt that glyptodons were as big as a car though, with the biggest being more than 10 ft (3 m) long and weighing in excess of 2 tons (2 tonnes).

A glyptodon's shell would have given it protection from pretty much any predator, but this came at the cost of the energy required to lug it around. The animals had short, very stout legs, which would have helped them carry this additional weight, but definitely indicate that they weren't sprinters. The legs could be held underneath the shell in the event of an attack, but the head and tail could not retract so both of these were also armored. Since glyptodons were a bit like giant tortoises, you have to wonder why they had a tail at all—it's not like they needed one for balance in

most circumstances. The first clue comes from the fact that in some species the tail had a cluster of spikes at the end. It seems that the musculature and design of the tail allowed the animals to swing it as a weapon—a bit like a medieval mace. This would have helped them ward off predators, but they probably had all the protection they needed in the form of the shell and wouldn't often have had to go on the offensive. Likely, the main benefit of the tail as a weapon was against other glyptodons—perhaps in male–male competitions related to breeding, or perhaps simply to stop others pushing them off a particularly tasty feeding spot. My guess is that the first of these is more likely, as I don't believe such giants lived in tightly packed herds where they were tramping on each other's toes when feeding; and one of the advantages of large size is that you don't have to be a really picky eater.

As with many other large animals that went extinct in the last 50,000 years after living for a million years or more, I think it is more likely than not that humans where involved in the demise of the glyptodons. In addition to being a source of meat, the animals had a large shell that some people have suggested would have made a durable shelter for humans to live in.

Meters

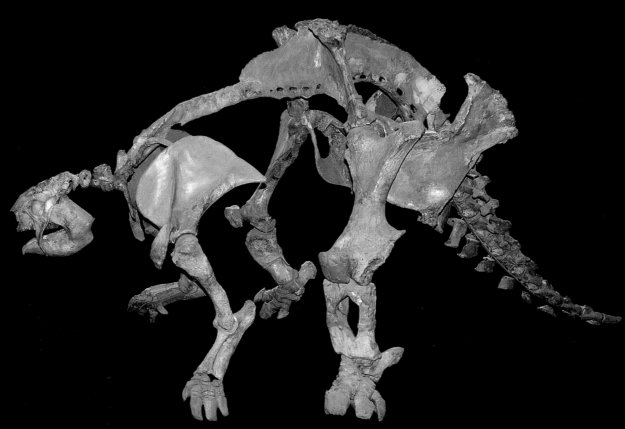

Gigantic Ground Sloths

We think of sloths as very slow-moving arboreal leaf eaters, but between 5 million and 10,000 years ago, South America was home to giant sloths that lived on the ground. The group that is of interest to us is the genus *Megatherium*, because some of these species were massive, measuring 20 ft (6 m) from head to tail and weighing 4 tons (4 tonnes). What makes them pretty unusual, however, is that they seem to have had very good balance as they apparently reared up on their hind legs to feed on trees that would have been inaccessible to other terrestrial grazers. They had a long, broad, muscular tail that helped with balance when they rose up, allowing them to rest on a tripod of their rear legs and their tail. They had really long claws on their front feet that would have been useful for pulling branches down toward their mouth. In fact, these claws were so long that a *Megatherium* would

Megatherium
Length: up to 20 ft (6 m)
The biggest of these ground sloths were the size of modern elephants and among the largest land animals ever. Like all the biggest mammals, they were herbivores. However, with their massive size and huge claws at the end of strong arms, they were certainly able to defend themselves against predators.

have had to walk on the sides of its feet to avoid putting too much weight on its claws alone—just like some extant anteaters. We have found trackways that suggest that these animals could sometimes walk just on their hind legs. My guess is that if they were foraging, then remaining bipedal as they changed position would have saved the effort of raising and lowering the body, but that quadrupedal walking would have been faster and more energetically efficient for moving when not feeding.

Elephants have very cushioned feet to absorb the shock when each is planted on the ground and a portion of the weight of the animal is brought to bear on it. This explains why I think a giant ground sloth must have had superb balance. When it had finished rearing up for food, my guess is that it would have been able to use its tail as a counterweight to lower itself onto its front feet slowly and gently. I think bringing its feet down with a crash repeatedly would have put too much strain on the front legs and led to frequent injury. Even putting the legs down gingerly would still have involved considerable strain, as they would have gone from being under no pressure to being asked to sustain a large proportion of the 4-ton (4-tonne) mass of the animal. Minimizing this strain by remaining bipedal for sustained periods while feeding would therefore have helped guard against repetitive stress injuries.

Again, I do think it likely that humans were involved in the demise of the giant ground sloths, but I don't rule out climate change as a contributing factor. These animals would have been slow breeders and a lot of space would have been necessary to support a viable population—as is the case with modern elephants. Any reduction in land area available to them, coupled with an increase in mortality through human hunting, might have been enough to drive them to extinction. To me, they look like they may have been easier to kill than glyptodons, since their hide could probably have been penetrated at any point by a robust spear. Given the challenge of walking on the side of the front feet, as discussed above, the sloths almost certainly wouldn't have been able to move as fast as modern elephants, and their size and the sound of them pulling down large branches would have made them easy to find and hard to lose. My guess is that, once they located a giant ground sloth, a band of humans with robust spears could have charged its flanks and done sufficient damage in a single attack that they could then simply have waited at a safe distance for blood loss to take its inevitable toll.

Mutual Benefits

Modern-day sloths do everything in the trees—except defecate. Once a week, they expend energy and risk predators by climbing down to the ground to poop. The reason for this is apparently linked to sloth moths, which are found in sloth feces only as larvae and in sloth fur only as adults. When the sloth reaches the ground, it digs a hole and poops in it, at which point female moths will emerge from its fur and lay their eggs in the fresh deposit before the sloth covers it over. The caterpillars that hatch out have this food source all to themselves and complete their development underground. The adult moths then fly up into the canopy to look for a sloth to hide in, thus completing the life cycle.

If the sloth simply defecated in the trees, the moths would find it much harder to find somewhere to lay their eggs and their offspring would face more competition from dung beetles and the like, so the sloth is doing them a favor by climbing down. In return, the moths do sloths a favor. Algae grow on the fur of sloths, often making them look stained green, and the sloths harvest and eat this. The moths, through their own feces and their dead bodies, fertilize the fur and promote algal growth. Thus, sloths with moths in their fur also have more algae to supplement their leaf diet. The moths and sloths have a very unusual mutualism.

▼ The ground sloths are extinct, but their tree-living ancestors are still widespread across South America.

Polar Bears

As we saw in Chapter 1, the largest animals tend to be herbivores rather than carnivores, which is why we have discussed vegetarians exclusively in this chapter so far. However, we will now turn to look at some really large mammalian carnivores. We will save the crocodiles and large snakes for our chapter on reptiles, and the toothed whales and seals for the chapter on aquatic giants, and instead here focus on bears, before turning to the big cats. We start by considering the furry white giant of the Arctic— the polar bear (*Ursus maritimus*).

The Polar Bear

Between 20,000 and 30,000 polar bears are widely scattered within the Arctic Circle. Polar bears and brown bears are very closely related. They occasionally interbreed, when unusually mild summer weather causes brown bears to venture further north, just as polar bears venture south in search of alternate food if the seasonal ice on which they normally hunt has not formed fully enough. This interbreeding makes it hard to work out (by comparing genetic samples) exactly when brown bears and polar bears separated from each other. Indeed, some brown bears might be more genetically similar to some polar bears than they are to some other brown bears, so some people would argue that you can't really call them different species. However, they are sufficiently different in color, foraging behavior, and distribution that I am comfortable considering them separately.

Polar bear
Ursus maritimus
Weight: up to 1.1 tons (1 tonne)
Polar bears prefer to prey on seals, catching them as they come up for air or haul themselves onto the ice.

◀ Polar bears are at home in the water and are capable of swimming long distances.

▼ The large paws of polar bears act as paddles when they are swimming and as snowshoes on land.

mounted rearing on its hind legs. Somewhere between 770 lb (350 kg) and 990 lb (450 kg) is normal for a male, although 1,300 lb (600 kg) would not be considered highly unusual. As mentioned above, polar bears prefer to eat seals and their digestion seems to be attuned to processing blubber effectively, to the extent that they often eat only the blubber on a seal and leave the meat. They catch seals in a number of ways, but all involve ice.

Seals often feed under parts of the Arctic ice that form seasonally, where the waters are both calm and full of food. However, they need to return to a hole in the ice to breathe periodically. A polar bear will wait by a commonly used hole, detecting a surfacing seal either by the sound or the smell of its breath, and will then swat it with a giant paw or dive in fully to catch it. Seals sometimes haul themselves out onto the surface of the ice to rest, either near the edge of an ice floe or near a hole large enough that they can dive back into the water at the first sign of an approaching bear. The bears are camouflaged against the ice and snow, and will patiently and very slowly try to creep up on a seal undetected, before making a final dash in the hope of catching it before it dives back into the water. Seals also give birth out of the water on a shelf that the female carves from the snow, sometimes meters deep. Polar bears can sometimes detect these by smell and sound, and then attempt to smash their way through the snow roof of the lair. It is said that the bears can smell a lair buried 5 ft (1.5 m) under the snow from a distance of more than half a mile (1 km).

One way polar bears and brown bears differ is in whether they think of you as a meal or not! Both of these species do kill humans occasionally, but the circumstances vary. Brown bears are territorial and may attack anything large that they feel is infringing on their property. You can also imagine that a female would attack if she felt her cubs were in danger. A brown bear might also attack if it is attracted to human food and is then disturbed eating it, especially if it feels cornered—although as soon as it sees an escape route, it will likely bolt and not give you a second thought. Polar bears are different. Seals are their preferred prey, but if they are unavailable then hungry polar bears will turn their attention to anything they can find to eat—including humans, which they will actively hunt in a way that brown bears would not. For this reason, most attacks on humans by polar bears are fatal, because the animal's motive is to kill and eat its prey. If it's any comfort, polar bears hunt by stealth, so if one did attack you, the first you would know about it would probably be when it made physical contact with you, after which death is likely to ensue mercifully quickly. As very few people venture into the part of the world in which polar bears live, very few are actually killed by them (only two in North America in the last 30 years).

Polar bears are big, and males are bigger than females. A male shot in Alaska in 1960 weighed 2,209 lb (1,002 kg) and stood 11 ft 2 in (3.39 m) tall when

Other Bears

The brown bear is another giant land-based predator, but unlike polar bears it has a more typical bear ecology. The North American short-faced bear (*Arctodus simus*) and Eurasian cave bear (*Ursus spelaeus*) were even larger, but both are now extinct.

Kodiak bear
Ursus arctos middendorffi
Weight: up to 1,660 lb (720 kg)

This is one of the largest subspecies of brown bear, and is confined to the Kodiak Islands off Alaska. In North America the brown bear is generally called the Grizzly.

The Brown Bear

As we saw on the previous pages, the brown bear is very similar to the polar bear in many regards, including size. However, it is very different in terms of distribution. Brown bears can be found in a huge variety of habitats, from estuaries to mountain forests, and are widely distributed in nearly fifty countries across North America, Asia, and Europe. Because of this, numerous subspecies are recognized, which can vary in their typical size. Generally speaking, the largest brown bears are of the Kodiak bear subspecies (*Ursus arctos middendorffi*), with a population numbering around 3,500 and living on the Kodiak Islands just south of mainland Alaska. The largest wild individual recorded was a 1,658 lb (752 kg) male shot in 1894. A captive animal called Clyde died in 1987 at Dakota Zoo in Bismarck, North Dakota, when he weighed 2,130 lb (966 kg).

Brown bears are particularly good at digging thanks to their claws, and soil invertebrates, burrowing animals, and bulbs and tubers can be an important part of their diet. They will also eat a wide variety of vegetation and fungi, especially different fruits. Scavenging from dead animals can be another important source of nutrition, and they hunt a very wide variety of prey, from small rodents all the way up to large deer—and everything in between. The hallmark of the brown bear is dietary flexibility. This is likely why they are not of strong conservation concern, which is very unusual for a large carnivore.

◀ Like most bears, brown bears (*Ursus arctos*) are extremely flexible foragers, sometimes living on berries, sometimes scavenging wolf kills, and sometimes fishing—like this one.

Cave bear
Ursus spelaeus
Weight: perhaps up to 2,100 lb (950 kg)

One of the largest extinct bears, the cave bear was probably similar in size to the very biggest of the polar bears living today.

There are a few brown bear attacks on people each year, and these are often fatal. Unlike the polar bear, the brown bear does not generally target humans as food. Rather, attacks generally result from the bear (often a female with cubs) being surprised by a human and feeling unable to escape. The best way to avoid being attacked by a brown bear is to wear bright clothing and make lots of noise, so it gets plenty of warning that you are nearby and can move out of your way.

Brown bears also come into conflict with livestock farmers, because they see sheep and other farm animals as an attractive source of food. Traditionally, keeping a large dog with the sheep can be enough to persuade bears that there are easier ways to get a meal. It's no surprise that such flexible feeders can also take an interest in human refuse as a potential source of food, and this can be a cause of nuisance and alarm. Thankfully, this is a relatively uncommon activity, since generally the bears are reluctant to interact with people. In North America, the much smaller black bear (*Ursus americanus*) is generally more willing to approach human habitation than the brown bear, which explains why it is responsible for about as many human fatalities in the US as that species.

Extinct Giant Bears

The short-faced bear was widespread across North America from about 800,000 years ago until about 11,000 years ago. Remains suggest that, on average, it was larger than the modern-day polar bear and Kodiak bear, with a weight of up to 2,100 lb (950 kg). The largest individuals could look me square in the eye when on all fours and rear up to perhaps 12 ft (3.7 m). Scratch marks on a cave wall that seem to have been

made by one of these giants reach up 15 ft (4.6 m) above the floor. As far as we can tell, the short-faced bear had an ecology very much like that of today's adaptable brown bears.

The cave bear lived in Europe and Asia until around 24,000 years ago, and might have reached similar maximum sizes and had a relatively similar ecology to the short-faced bear. The reasons for its extinction is unclear, and persecution by people seems unlikely to be the main cause in this case because human populations were pretty small across its range at the time, and because it is very rarely depicted in cave paintings. The species' common name stems from the fact that remains are often found in caves, suggesting that these were often used for shelter. It may be that these relatively scarce resources were soon monopolized by early humans and Neanderthals, and that this was enough to tip the cave bear into extinction, but that doesn't feel entirely convincing to me and it doesn't chime with their rarity in cave art.

Pure Predators

The big bears can be lethal to humans and even to animals a lot larger than us, but the real key to their success is flexibility—fruits and roots matter just as much to a brown bear as does hunting. In terms of pure predators, the biggest are the dogs, cats, and hyenas. In the case of dogs and hyenas, they don't have to be very large at all in order to tackle big prey because they rely on teamwork. However, with the exception of the lion, hunting is normally a solitary activity for the cats. The biggest of the cats is the tiger (*Panthera tigris*), which uses its enormous power to hunt prey my size or bigger. It really is a terrifying predator.

The Tiger

Of the tiger subspecies, the Bengal tiger (*Panthera tigris tigris*) is generally considered to be the largest, with typical adult males weighing close to 440 lb (200 kg). Hunting tigers was a common sport through the last nineteenth and twentieth centuries, so we have a lot of data on the sizes of the largest individuals shot. The record-holder is a Bengal tiger that was killed in 1967 and weighed 858 lb (389 kg). There isn't much difference in size between tigers and lions, but it seems that tigers are a little bigger on average. It was once fashionable in zoos to breed the two species, and the hybrids produced from male lions and female tigers (ligers) are much larger than any wild individual of either species, weighing up to 1,100 lb (500 kg) when fully grown.

▼ Hybrids of lions (*Panthera leo*) and tigers (*P. tigris*) were once popular in zoos, largely because of the huge size they could reach (aided by abundant food).

Tigers hunt by stealth, targeting animals that weigh about 200 lb (90 kg). When food is plentiful, they prefer to pass up small prey and instead wait for something big—equal to their own size or more. They seem fearless, and with the element of surprise will attack other large, fierce predators, including brown bears (see page 60), leopards (*Panthera pardus*), hyenas, wolves (*Canis lupus*), large snakes, and crocodiles. Part of the reason for this behavior is that tigers are very territorial and are keen to cut down competition for prey on their patch. Their method of hunting by stealth leads to territoriality—they need to know the local area intimately and where potential prey might pass handy hiding places, but they need a territory that is large enough such that potential prey can't predict where they will be.

Given our recent history of persecuting tigers, most stay clear of humans. However, a small minority specifically target people as prey. These are often sick or injured individuals that aren't strong enough or agile enough to take down the types of prey they normally prefer. However, deliberate targeting of people is not entirely confined to weakened tigers (or lions). Human activity in an area tends to reduce the availability of large wild animals as potential prey, and this can tilt the balance in favor of deliberate hunting of humans. It certainly does seem that some individuals become man-eaters, as attacks on humans in a given area can be unusually high until a particular cat is shot. This has less to do with acquiring a taste for humans, and more to do with having learned profitable hunting locations and techniques.

Tigers have been in spectacular decline. One issue is human-caused mortality, either for sport, to service demand for traditional medicine, or because of real or perceived danger to people and their livestock. Another challenge is the species' extreme habitat

Tiger
Panthera tigris
Weight: at least 675 lb (300 kg)

A tiger's long, strong canines can puncture the hide of any animal it attacks.

▲ Tigers generally specialize in large-bodied prey like this unfortunate deer.

▼ Hunting tigers for sport has long been fashionable among the ruling classes across the species' range.

requirement—each tiger needs a large territory (at least 8 square miles, or 20 km²) stocked with big game, with suitable vegetation to provide cover, and containing fresh water and somewhere to den up and rest. Human encroachment is making such areas harder and harder to come by. It is estimated that the species' range has contracted by 93 percent over the last 100 years and that their numbers have fallen from 100,000 to between 3,000 and 4,000 over the same timescale. Worse still, this population is fragmented, and it is considered that no subpopulation has more than 250 reproductive individuals and that some have a lot less—there are probably more tigers in the world's zoos today than there are in the wild. However, the iconic status of tigers has led to considerable investment in their conservation, and there is some hope that their numbers might be stabilizing (but still at a perilously low level). We need to maintain that investment—it would be tragic if this majestic giant were to become extinct on our watch.

Lion
Panthera leo
Weight: up to 550 lb (250 kg)
Life is seldom quiet for the dominant male in a pride—his position is regularly challenged by other males.

The Lion

Much of what we have just read about tigers pretty much holds for lions too. The big difference between the two species is sociality. Lion prides can be very variable in size, ranging from three adults all the way up to twenty, and they can also vary in their sex ratio, although this is generally heavily skewed toward females. Prides usually hold a defined territory, but there are also (often young or very old male) lions that lead a more nomadic existence on their own or in a pair.

In terms of feeding, the obvious cost of sociality is that kills have to be shared. This is particularly painful if the male doesn't pull his weight on the hunt but still expects first refusal on any kills. Females often live with this because the male defends the cubs against other males—the cubs are likely his offspring, so he is invested in defending them. If he were usurped by another male, then this usurper might well kill the cubs since they are unrelated to him and he would rather the females focused their efforts on reproducing with him instead.

The benefits of sociality are that, by working as a team, it should be possible to capture prey that would be too fleet or too strong for a single individual. In addition, there is a certain amount of insurance in this: a lame tiger simply does not eat, whereas a lame lion in a pride will likely be allowed to share in kills for a while until it recovers. Sociality in lions also means that they don't have such extreme habitat requirements. Both tigers and lions ambush their prey, but tigers need cover that allows them to get very close to their prey. This requirement is relaxed in lions, since if the target animal flees one lion that it has detected, it may well run right into the jaws of another. Finally, being in a group makes it easier to defend kills against hyenas and other potential scavengers.

What Big Teeth You Have!

Extinct giant cats were never much bigger than existing lions and tigers—perhaps 10–25 percent bigger in the American lion (*Panthera leo atrox*) and Eurasian cave lion (*P. spelaean*). As we have seen, modern-day lions and tigers need huge home ranges to have enough of a prey base to support their food requirements, and from Chapter 1 we know why large predators live at very low population densities, so there would have been little incentive for the ancestors of these big cats to grow much larger, except to allow predation of really large prey. There were generally more megaherbivores in the recent past than there are today but, as we have discussed (see page 20), these animals would themselves have lived at low densities. And just as modern tigers and lions can manage to take down impressively large prey, it appears that there isn't sufficient advantage in any feline evolving to be much bigger than today's big cats.

One trait that has evolved separately in predatory mammals at least seven times is huge canines. The most famous group of sabertooths is the genus *Smilodon*, of which the South American *S. populator* was probably the biggest at weights that may have reached 880 lb (400 kg) and canines measuring 11 in (28 cm) in length. Such teeth can potentially break if they strike bone, and would generally be useless and even get in the way when trying to eat small prey. It is therefore assumed that saber-toothed cats were specialists on large prey. They usually had a huge gape that allowed them to direct stabbing actions of their teeth deep into the neck of a victim while using their notably strong forelimbs to hold the prey down and allow accurate biting. We don't see sabertooths today because the last 25,000 years have seen a sharp reduction in large animals, although I will resist repeating (again) why I think that is!

▲ Saber teeth seem ideal for inflicting terrible damage on even the very largest prey.

◀ Teamwork allows lions to bring down prey larger than an individual could manage alone, but it does mean that meals must be shared.

The Forgotten Aquatic Mammals

If I asked you to name an aquatic mammal that never leaves the water, you would probably think of a whale or a dolphin, which we cover in the next chapter. But there is another group of completely unrelated mammals that have returned to an obligate existence in the water—the Sirenia, or sea cows, which are the closest living relatives to elephants.

Living Sirenians

The living sirenians are the dugong (*Dugong dugon*) and three species of manatee (genus *Trichechus*), all of which are herbivorous mammals that live in rivers, swamps, marshes, and coastal regions. They can grow to 13 ft (4 m) in length and weigh up to 3,300 lb (1,500 kg). Their herbivorous diet is quite different from the carnivory exhibited by all the whales, dolphins, seals, and sealions. Just like whales and dolphins, they have evolved to spend their whole lives in the water and never venture onto land. Unlike the whales and dolphins, however, we have a pretty clear idea how these mammals evolved through intermediate stages from land mammals to species that now have entirely aquatic lifestyles. Fifty million years ago, terrestrial ancestors of the modern sirenians that fed in marshy areas gradually became increasingly specialized for swimming to feed on freshwater plants, until by 40 million years ago they were fully aquatic.

It is no surprise that sirenians today inhabit shallow waters, as this is where aquatic vegetation flourishes (deep waters are too dark for photosynthesis). Even though they feed on the most nutritious vegetation, the plants they select are still tough and fibrous, and are low in energy compared with the carnivorous diet of whales and dolphins, so they have to spend a lot of time feeding. Being mammals, this is a problem for the sirenians, because they have to surface to breathe. They combat this by having a low metabolism, so they don't have to breathe too often, and by feeding in as shallow water as possible, requiring them simply to raise their head to snatch a breath before quickly returning to feed. They also have the densest bones of any mammal, allowing them to rest on the bottom without having to expend energy to stop themselves from continually bobbing up. Given their slow metabolism, it's no great surprise that sirenians favor warm waters, although until recently one of their number did live in cold waters.

Dugong
Dugong dugon
Weight: up to 1.7 tons (1.5 tonnes)

Note the extensive scarring on this individual—dugongs often feed in relatively shallow waters close to humans, and entanglement in fishing nets and vessel strikes are common hazards.

West Indian manatee
Trichechus manatus
Weight: up to 1.8 tons (1.6 tonnes)

The slow movement and docile (even inquisitive)
nature of manatees often allow divers to get
spectacular close-up views of these giants.

The Individual Species

The West Indian manatee (*Trichechus manatus*) can
be found throughout the Caribbean and a little way
up the eastern coast of the US. In recent decades,
individuals have congregated around the outflows
of nuclear power stations in winter, where they can
be bathed in the water used to cool the reactors.
Unfortunately for the manatees, these power stations
are coming to the end of their working lives and are
now being replaced by more efficient plants that
produce nothing like the same waste heat and so
won't be nearly as attractive to the sirenians. Although
some scientists are worried that the manatees may
have become hooked on these winter refuges, my
guess is that they will adapt without difficulty to their
loss by returning to migrating farther south in the
winter. These are long-lived animals, reaching perhaps
sixty years of age, so nuclear power stations have
been around for only a few of their generations and
their previous natural behavior is unlikely to have been
completely purged from the population.

The Amazonian manatee (*Trichechus inunguis*)
lives entirely within the Amazonian river system,
which experiences dramatic seasonal expansion and
contraction associated with a defined rainy season in
the catchment hills to the west. This means that some
individuals end up spending the dry part of the year
in a small landlocked lake with little to eat. Thanks to
their low metabolism, these beasts can survive on their
fat stores until they can escape into the wider river
system once water levels rise again.

A third manatee species, the African manatee
(*Trichechus senegalensis*), lives in river estuaries and
coastal systems along the western coast of Africa.
Like all sirenians, it shares its habitat with some
potent predators (most obviously crocodiles and
sharks), but predation on adults seems very low to
nonexistent. This must be put down to the large size of
the manatees and also perhaps their ability to deliver
powerful blows with their tail—it is easier to imagine
predation on juveniles, but this is yet to be investigated
in any detail.

The dugong is scattered across the coast of the
Indian Ocean and the western fringes of the Pacific;
it is never found in rivers and is entirely marine
throughout its range. Research on dugongs suggests
that these sirenians are not as strictly herbivorous as
originally thought. They are definitely predominantly
herbivorous, but they will consume invertebrates
when they get a chance and will even steal fish from
fishermen's nets.

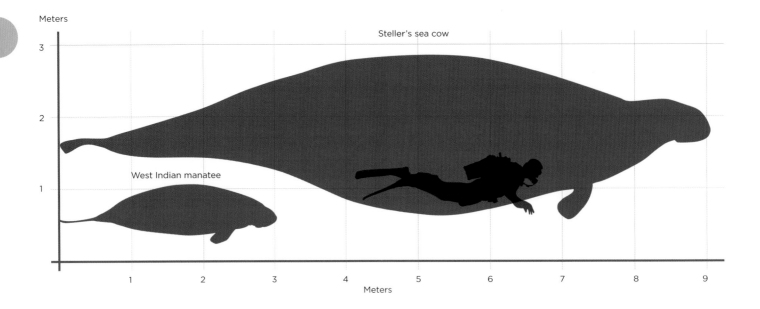

Meters

Steller's sea cow

West Indian manatee

Meters

Vulnerable Status

As you might expect, the coastal, estuarine, and riverine preferences of the sirenians bring them into close contact with humans. All the species have been hunted for their meat, their oil, their bones (for medicinal use), and their thick hide (for clothing and shoes). They are slow-moving and generally inquisitive, and so are very easy to catch. These traits also make them vulnerable to boat strikes, which are a significant source of injury and mortality for West Indian manatees in particular, and their habitats are vulnerable to coastal development and pollution. They can be found in a scattering of zoos around the world, but are neither the easiest animals to keep nor a huge draw for visitors. The best way to see these giants is to go snorkeling in areas where they occur naturally. They may be inquisitive, but they have never shown any aggression toward humans in their vicinity.

Steller's sea cow
Hydrodamalis gigas
Weight: perhaps up to 11 tons (10 tonnes)

As with its living kin, Steller's sea cow had a snout that points downward to aid gripping and pulling seaweed from the substrate.

A Steller Sirenian

Steller's sea cow (*Hydrodamalis gigas*) was discovered in 1741 and extinct by 1768. It was the largest sirenian known, reaching up to 30 ft (9 m) in length and weighting around 9–11 tons (8–10 tonnes). Completely unlike its living cousins, it was a cold-water specialist that was confined to the sea off the Commander Islands between Russia and Alaska. Subsequent discoveries of bones suggest that the species once had a broader distribution, but scientists have little idea how far this extended and what led to its concentration around the Commander Islands.

The large size of Steller's sea cow can be seen as an adaptation to the cold waters in which it lived, as it reduced the animal's surface area to volume ratio. Heat loss was further minimized by a relatively thick blubber layer, which also gave the sea cow positive buoyancy, allowing it to feed on blades of kelp at or near the surface. A slow-moving surface feeder would have been easy for sailors to catch, and it is also said that the meat kept fresh for an unusually long time, likely due to the high salt levels retained from the kelp on which the animal fed. It has been estimated that there were about 1,500 Steller's sea cows at the time of their discovery, and news of this easy means of replenishing ships' supplies in a generally barren part of the world traveled fast—the result being that they were all gone within a short space of time. Today, all that remains are about thirty near-complete skeletons and seventy skulls in various museums, and the original report on the species by German naturalist Georg Steller—including descriptions of its anatomy, behavior, and tastiness (see box).

Steller and Bering

Vitus Bering (1681–1741) was a Danish-born officer and explorer in the Russian navy. He took the German naturalist Georg Steller (1709–56) with him on a 1741 voyage to explore the geography of the region between the Asian and American continents. Steller was both scientist and doctor for the expedition. Things did not go well on the return home, however, and their ship foundered in a storm in what was later called the Bering Sea. Many of the men perished, including Bering himself. Those remaining ascertained that they were, in fact, on an island—later called Bering Island—and so the chance of walking out was zero and the chance of being discovered by passersby was similarly slight. They eventually fashioned a new vessel from the remains of the old and sailed back home.

During the castaways' several months on the island, Steller took meticulous notes of the wildlife, including the only recorded observations we have of Steller's sea cow. The shipwrecked sailors were the first nonindigenous people to hunt the animal, and this was likely a critical factor in the survival of the men, who were already in a weakened state after a long voyage. Following their return, news of the islands in the Bering Sea and their food supply was of real interest to sailors using these waters, most of whom were involved in the harvest and trade in furs.

Meanwhile, Steller travelled extensively along the very eastern fridge of the Asian continent and studied the indigenous peoples there. His concern at their treatment by the Russian state made him unpopular with the government, and he died soon after narrowly escaping a jail sentence for fomenting insurrection. This likely explains why Bering's name appears on maps more often than that of Steller. Steller's nature observations were published posthumously, and a jay, a sea eagle, a sea lion, and an eider duck have all been named after him in addition to the sea cow. Steller also discovered the largest cormorant known: the spectacled cormorant or Pallas's cormorant (*Phalacrocorax perspicillatus*) on Bering Island. Described as flightless or near-flightless and "good to eat," it ultimately—and not surprisingly—suffered the same fate as the sea cow and on a pretty similar timescale.

▼ Scientists can make a good guess at what Steller's sea cow looked like in life from the thirty or so complete skeletons that have been found, as well as the general morphological similarity among its living relatives.

Chapter 4
GIANTS OF THE DEEP

In Chapter 1 we saw how the buoyancy offered by water greatly reduces the cost of supporting a bigger mass, so it shouldn't be a great surprise that the oceans are good places to look for giants. The blue whale isn't just the biggest animal in the world today; as far as we know, it's the biggest animal that has ever lived. But there are some breathtakingly big fish too, and the largest living reptiles are also swimmers—although here we focus on their even bigger ancestors, saving modern-day marine turtles and crocodiles for Chapter 8.

Gentle Giants

Sharks have a fearsome reputation, but the biggest species mean us no harm at all. The largest members of this group aren't really predators in the true sense, but instead filter great quantities of water, sieving out tiny creatures and fish eggs. While it makes sense that predators should be big enough to overpower their prey, but not so big that the prey can outmaneuver them, there is no restriction on size for filter feeders relative to food particles. Indeed, large size confers benefits, such as deterring would-be predators.

A Whale of a Fish

The whale shark (*Rhincodon typus*) has cruised the warmer oceans of the world for 60 million years. These gentle giants swim slowly along with their huge mouth gaping open, pushing in great volumes of water that then exit via long gill slits around where you'd imagine the neck to be. All fish have gills, which are the organs that allow oxygen to be absorbed from the water. Commonly, there is a protective apparatus before the gills, which sieves out any material in the water that might damage these fragile structures. Hence, it is very easy to imagine how these structures have been selected by evolution to be particularly efficient at collecting almost any suspended material. In clear seas, almost everything in the water is living material worthy of eating. Every so often, the material that builds up on the filters is shaken off with something like a cough and swallowed. If the whale shark is in a particularly rich patch of food, then it doesn't even have to swim to collect a meal—it can just repeatedly suck in great volumes of water as it sits stationary.

The water that is sucked in is then replaced by surrounding fresh seawater, along with a fresh complement of food.

Whale sharks are massive. The biggest reliably measured individual had a length of 41.5 ft (12.6 m, or about the length of three compact family cars parked nose to tail) and weighed 23.7 tons (21.5 tonnes, or a bit more than fifteen compact cars). These benevolent behemoths can crop up anywhere in warm equatorial waters, although they mostly seem to stay offshore. If you do come across one, then note that youngsters are known to take an interest in divers without ever harming them, and larger individuals are generally oblivious even when divers approach very closely.

Basking in Glory

The basking shark (*Cetorhinus maximus*) is the world's second-largest fish. It is broadly similar in morphology and lifestyle to the whale shark, yet is not a close relative. The two species have divided up the oceans between them, with the whale sharks sticking to warm waters near the equator, and basking sharks remaining in cooler temperate waters. The largest basking shark was found by some herring fisherman when they dragged their catch onto their boat off Canada in 1851—it was 40 ft (12.3 m) long and weighed 17.6 tons (16 tonnes). Sadly, and very tellingly, few individuals exceeding 26 ft (8 m) have been found in recent years.

The basking shark has long been hunted for both its oil and its skin. It was an attractive target species because it was once relatively commonplace, it is placid and slow-moving, and it favors both surface and inshore waters. Hunting for skin and oil has now largely ceased, but the fish is still sought after as an ingredient for shark's-fin soup and for traditional medicine in parts of Asia. Even

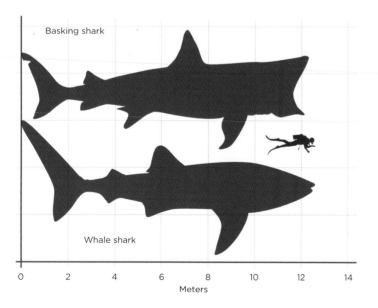

Basking shark

Whale shark

0   2   4   6   8   10   12   14
Meters

with the reduced levels of hunting, it remains vulnerable to being accidently caught in fishing nets, since the same productive waters that attract it are likely to appeal to some of the shoaling fish targeted by commercial fishermen. Being slow-moving and willing to feed in estuaries and even harbors, the basking shark is also relatively vulnerable to boat strike. However, not all is gloom and doom. The very qualities that once made the shark an attractive target species also make it appealing as an ecotourism drawcard, which may provide an added financial incentive to conserve the species.

The basking shark is a very simple giant. Unlike the whale shark, it can't actively pump water into its mouth and so must always be swimming forward to feed. It has an extraordinarily small brain for its size, but much of that can be put down to its simple lifestyle. That said, satellite tagging has revealed that these animals are adept at seeking out the most profitable parts of the ocean year-round, which often requires them to cover huge distances. Although they spend much of their time alone, they congregate in productive waters in summer in particular, and in this situation they display social behaviors. So, while basking sharks might be small-brained, they live rich lives nonetheless.

Whale shark
Rhincodon typus
Length: up to 41.5 ft (12.6 m)

This whale shark has been lured to investigate a net full of fish near the surface, providing a perfect photo opportunity for the divers.

Basking shark
Cetorhinus maximus
Length: up to 40 ft (12.3 m)

Basking sharks favor upper water layers near coasts, swim slowly, and are very docile, making them relatively easy to photograph.

Megamouth

It is worth mentioning here another giant shark filter feeder, the megamouth shark (*Megachasma pelagios*). It's a little smaller than the other two species and is only distantly related to either of them, but it has a broadly similar anatomy and lifestyle. Again, it has carved out its own part of the oceans—whereas the whale and basking sharks filter surface waters, the megamouth lives lower in the water column. This explains why a leviathan that can grow longer than 16 ft (5 m) and weigh more than 2,000 lb (900 kg) wasn't discovered until an individual became tangled in the anchor chain of a US warship in 1976. That shark was 15 ft (4.5 m) long and weighed 1,650 lb (750 kg), and since then, just over fifty have been inadvertently caught or sighted. As we will see in Chapter 9, photosynthesis occurs only in the top layer of the ocean and so there is less food to go around in deeper waters. For this reason, the megamouth is very focused on conserving energy—it swims at just 1.2 mph (2 kph), or slower than the average human walking pace. What it lacks in speed it makes up for in other features, however, in that it has a number of light-producing organs around its mouth. It is possible that small prey actively swim to their doom after being attracted by these lights, but this (and much of the biology of this marine monster) has yet to be confirmed.

Record-holding Ray

The group most closely related to sharks are the rays, and the largest of them, the manta rays, feed on tiny organisms that drift through the oceans. Their approach is a little different from that of the giant sharks discussed so far, which should be no surprise given their radically different body plan. They are also filter feeders, but they often herd prey together by swimming in lazy circles around a volume of water. The currents produced by the huge outstretched wings bunch fish together, and once the ray has a tight ball of prey, it zooms through it at high speed, hoovering the fish up before they have the chance to drift away. The giant oceanic manta ray (*Manta birostris*) is the larger of the two members of the genus, reaching 23 ft (7 m) across, about 10 ft (3 m) from front to back, and weighing about 3,000 lb (1,350 kg).

Megamouth shark
Megachasma pelagios
Length: up to 16 ft (5 m)
Photographs of this giant are very rare, as it lives in deep, dark waters.

Giant oceanic manta ray
Manta birostris
Width: 23 ft (7 m)

Like many giant fish, this ray has attracted an entourage of smaller fish; by following the ray, they gain protection from predators as well as easy food if the manta is a messy eater.

◄ Manta rays are often found in groups, as they can more effectively corral prey into a dense ball by working together.

Filter Feeders of the Past

There is a long history of filter-feeding giants. By the age of the dinosaurs, it appears there were large filter-feeding sharks and bony fish, although none of those found so far were as big as the giants we see today. We know about extinct filter-feeding sharks for a rather odd reason. All the extant large filter feeders described here evolved independently from ancestral predators. With the switch to filter feeding, the sharks had no need for teeth, so invested a lot less in them. The teeth have not been eliminated entirely by natural selection, but they are much smaller and much less refined than those of predatory sharks. All sharks have teeth that last only a few months, before these fall out and are replaced (this is why a shark's mouth seems to be a strange mess of different-looking teeth), so we have a fabulous fossil record of the teeth of prehistoric species. This include small, undefined teeth like those found in the mouths of filter feeders but not in any other sharks, so we know the former have been around for a long time.

The Blue Whale

Whales are separated into two major groups: the baleen whales and the toothed whales. The baleen whales sieve small food particles from the water, rather like the giant sharks discussed on the previous pages, while the toothed whales are predators on relatively bigger prey. Before considering these animals in general, we look at the biggest of them all—the blue whale (*Balaenoptera musculus*).

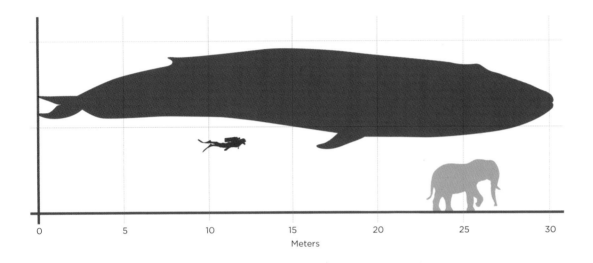

Meters

A Giant Among Giants

The blue whale is a record-breaker in lots of ways. First and foremost, it is the largest animal on the planet today, and as far as we know, it is also the largest animal that has ever lived. One (very small) compensation for the industrial-scale hunting of whales that occurred in the twentieth century is that we have a good idea of just how big the biggest blue whales can be—the longest measured by a scientist was 98 ft (30 m), but there are some reputable measurements that are even greater than this, including one at 110 ft (33.6 m). If you were to park a row of compact cars nose to tail again, eight would fit alongside the biggest blue whale.

Estimating the weight of a blue whale is a bit more tricky. Most sources quote a figure of 220 tons (200 tonnes, or 145 compact cars, in case you were curious), but this is a guesstimate for a number of reasons. First, blue whales carry out the bulk of their feeding in the highly productive polar waters in spring and summer (see box on page 79), so while a large individual might weigh around 150 tons (140 tonnes) at the start of the spring, by the end of the summer it will

have put on sufficient stores to see it through winter, increasing its weight by 50 percent to 230 tons (210 tonnes). Second, no apparatus is big enough to weigh a live, fully grown blue whale. Any attempts that have been made to weigh these animals have involved cutting them into chunks, weighing the individual parts, totaling these weights, and then adding an estimate for all the blood and body fluids lost in the process. The final complication is that the very longest whales measured have not actually been weighed. So, it's likely that there are, or have been, blue whales weighing more than 220 tons (200 tonnes), but probably none weighing as much as 250 tons (225 tonnes), and we might never get a more accurate estimate than that of the heaviest animal ever.

Multiple Record-breaker

The blue whale's records don't stop with its sheer size, as all the different parts of the animal are pretty astounding too. It is estimated that the largest individuals can hold 100 tons (90 tonnes) of water (and hopefully food) in their mouth at one time, which is equivalent to taking a bath a day for three months.

Blue whales need a huge tongue to lick the food from the curtains of baleen in their mouth, and this weighs 6,000 lb (2,700 kg), or thirty-four times the weight of a grown man. Its heart weighs 395 lb (180 kg), but the brain weighs less than 15 lb (7 kg). Sadly, the common claim that you could swim through a blue whale's major arteries is not true—they are big, but not that big. Finally, the blue whale is reputed to have the largest penis in the animal world, at 95–120 in (240–300 cm) long. This last appendage is difficult to measure, however, since it swells during sexual activity, as in humans, and any diver would be foolhardy indeed to attempt to get close with a tape measure when two 220-ton (200-tonne) giants were about to couple.

The upshot of such a coupling might be a calf some ten to twelve months later. At birth, a blue whale calf weighs 6,000 lb (2,700 kg), or as much as a fully grown hippopotamus (*Hippopotamus amphibius*), and will then put on as much as 200 lb (90 kg) a day, before weaning after six months having doubled in length. Human breast milk is 5 percent fat, 1 percent protein, and 7 percent carbohydrate, giving a calorific value of around 34 kcal/lb (315 J/kg). By comparison, a blue whale's milk is around 40 percent fat and 15 percent protein, adding up to 1,800 kcal/lb (3,440 J/kg). And the whale calf might drink about 100–150 gallons (400–600 liters) of milk a day, compared to just 2 pints (1 liter) consumed by a healthy human baby.

Blue whale
Balaenoptera musculus
Length: 98 ft (30 m)
Unlike sharks, whales must come to the surface to breathe.

▼ The exposed penis of a gray whale (*Eschrichtius robustus*). This is another giant baleen whale, with the largest individuals reaching 49 ft (15 m) in length.

A Near Thing

The blue whale is one of the speedier whale species, and this, combined with its size and power, meant that it was ignored by whalers—at least until whales that were easier to catch became rarer, harpoon guns became more powerful, and ships became bigger and faster. Hunting for blue whales began in earnest around 1880, and by 1945 numbers were very heavily depleted.

There is a huge amount of guesswork involved in estimating how many whales of any type there are in the world's oceans today, and even more speculation in trying to come up with a number for how many there were before whaling started. We don't even know how many whales were caught, because records were either not kept or were lost, and after restrictions to whaling were introduced progressively from 1945, records were falsified or whaling was carried out in secret. That said, it is generally considered that there were between 210,000 and 310,000 blue whales prior to whaling. Perhaps 95 percent of these were concentrated around Antarctica, and it was this concentration that the whalers specifically targeted—in the Antarctic summer of 1930–31 alone, records suggest that around 30,000 blue whales were caught. Today, there are probably between 10,000 and 25,000 blue whales globally. Given the terrifying scale of commercial whaling, it is amazing that humans didn't drive any species to extinction. This is testament to just how big the oceans are, and also the high cost of running a whaling operation. Smaller populations didn't attract the attention of whalers, and for the largest Antarctic population, whaling was no longer economic once catch numbers dropped below a certain level. Tragically, it was for this reason rather than public pressure for the implementation of conservation measures that whales survived.

But it was a close call: it is estimated that the largest blue whale population around Antarctica was reduced from at least 200,000 to less than 400 in the 1970s.

Bouncing Back

Following the ban on commercial whaling in 1966, the blue whale population has certainly increased. The species' only possible predator is the killer whale (*Orcinus orca*; see page 94), and these hunters most likely concentrate on easier prey. Ship strikes and becoming entangled in fishing gear are a concern for many whales, but the former is less of an issue for blue whales, as they are relatively fast swimmers. However, we know little about the potential adverse effects of sonar and other human-produced noises in the sea on whales in particular, and on sea life in general. A large mammal obviously has to eat a lot, and blue whales do most of their feeding in rich polar waters in summer (see box),

◀ A depiction of whaling during the era of sailing ships, when whales were attacked with harpoons thrown by hand from small open vessels. Such operations were very dangerous for the whalers and often led to prolonged suffering on the part of the whales.

▲ Adult and subadult Antarctic minke whales (*Balaenoptera bonaerensis*) being dragged aboard the *Nisshin Maru*. This is both the largest vessel and the flagship of Japan's whaling fleet, and thus is often the focus for antiwhaling protesters. The Antarctic minke is the second smallest of the baleen whales and the most abundant; as such, it is a focus of both whaling operations and whale watching.

so climate change may easily affect the productivity of the seas in ways that impact these marine giants.

It is relatively easy to appreciate the majesty of blue whales in person, as many museums and other institutions around the world display skeletons, life-sized models, and even stuffed specimens, and the bones of stranded animals are often arranged into long-lasting monuments. On the Isle of Lewis in Scotland, there is an archway alongside the main road that is made from the lower jaw of a blue whale that beached there in 1920.

▲ In the Faroe Islands, locals still hunt whales for food.

▼ Today, numbers of many formerly persecuted whale species are on the rise.

Productive Polar Waters

It might seem strange that polar seas are actually more productive (in other words, produce more biomass per unit area) than tropical seas. After all, we are used to the idea that warmer temperatures are good for growth, and we definitely know that sunlight is very much lacking at high latitudes in winter. However, in spring and early summer there can be absolutely spectacular productivity in polar seas.

The issue is nutrients, and nitrogen in particular. In warm weather, the surface water of the ocean heats up and there is very little movement between this layer and the deeper ocean because it has slightly lower density and so naturally rises. There is enough sunshine in this upper layer to drive photosynthesis, but organisms need more than just carbohydrates from photosynthesis in order to grow. To make proteins, for example, nitrogen is required. Plants get this from the soil, and photosynthetic phytoplankton get it from the sea around them. However, the nitrogen in the surface layer of the oceans is soon used up by growing organisms. When those organisms die, they generally sink into the depths, where they might be broken down and some of their chemical building blocks return to the water. So, in summer there is generally a warm, nitrogen-poor ocean layer sitting on top of a colder, nitrogen-rich layer. Photosynthesis is therefore restricted by a lack of nitrogen in the surface layer, and by a lack of light in deeper waters.

This is the situation year-round in the tropics, but in winter nearer the poles, surface water cools, which makes mixing of the different layers easier, a process that is helped further as more frequent storms also tend to churn the waters. The upshot of this is that in polar seas the upper layer has its nitrogen store replenished over the winter, although it is clearly too dark then for much photosynthesis to take place. But come the spring, there is plenty of light and plenty of nitrogen, and consequently plenty of photosynthesis, and the waters teem with life—this is often called a spring plankton bloom. This never happens in the tropics, where the upper layer stays permanently short of nitrogen, restricting the rate of new growth.

Baleen Behemoths

There are fifteen extant species of baleen whale, and all are really big. As we have seen, the blue whale is the biggest, but the fin whale (*Balaenoptera physalus*) is not far behind, growing to at least 85 ft (26 m) and maybe as much as 90 ft (27 m). It is closely related to the blue whale, having separated from that species a relatively recent 1.5 million years ago, and the two seem to interbreed from time to time. Even the smallest baleen whale, the pygmy right whale (*Caperea marginata*), can reach a length of 11 ft (3.5 m) and weigh 3.8 tons (3.5 tonnes). Thus, among the land animals, only the elephants are heavier than the most diminutive members of this group of giants.

Filter-feeding Facts

Ancestrally, the whales were generally carnivores that hunted substantial-sized prey and relied heavily on their teeth. Some of these animals started supplementing their diet by gulping down smaller prey with a volume of water, and then spitting the water out again, using their teeth and pursed lips to sieve out the food. As this form of food gathering became more important, the internal mouth morphology of these whales evolved to make the sieving process increasingly efficient, including the growth of curtains of fibrous baleen, made from keratin. This is a common natural material, being found in mammalian hair, claws, horns, and hoofs. Eventually, the whales became more specialized still for this form of feeding, and gave up standard predation with their teeth altogether. The baleen whales split from the toothed whales about 35 million years ago.

Two techniques are used by baleen whales during filter feeding—skimming and lunging—and most species specialize in one or the other. Skim feeding (as used by right whales, for example) involves swimming along with the mouth open. Water and suspended food are washed in, and then the water is pushed out of the mouth by the pressure created by the forward movement of the whale. As this happens, the water is propelled through the baleen and the food particles are retained. This is a clever way to capture really small prey that can do little to escape the cavernous mouth coming toward it.

Lunge feeding works better against slightly bigger prey that might be able to flee an oncoming skimmer. Lunge feeders circle the food to gather it into a tighter ball, or they focus on a group that is already naturally in a ball. Either way, they swim toward the ball quickly enough that the individual animals have no time to

Fin whale
Balaenoptera physalus
Length: 85 ft (26 m)

The fin whale's sleek, streamlined body often allowed it to outrun early whalers, but its numbers suffered dramatically from more industrialized whaling in the twentieth century.

▲ One secret to the success of the baleen whales is their huge mouths, shown clearly in this portrayal of social behavior—another feature common to members of the group.

react, and then at the last moment they open their mouth spectacularly thanks to pleats at the side of the head. The mouth can open such that it has larger volume than the total volume of the whale with its mouth closed, and engulfs the ball and surrounding water. The whale then sieves the water through its baleen and back out through its mouth, licks the food off the baleen, and swallows. Blue whales use this technique to feed on their favorite prey—small ⅜ in–¾ in-long (1–2 cm) crustaceans called krill.

Baleen whales generally feed in the polar waters, and do most of their grazing there when the biomass in these waters is highest, in spring and early summer (see box on page 79). They then often migrate to warmer waters in winter. This might be because waters at lower latitudes offer more food than the polar seas in winter, it could be that they give the whales respite from the attentions of killer whales (which are more prevalent nearer the poles), or it might possibly be that warmer, calmer waters are needed for giving birth. However, it is likely a complex combination of all three factors.

Why so big?

So, why did baleen whales get so big? This is actually quite a tough question to answer. While the baleen whales have always been pretty big, it is only relatively recently that they have become gigantic. Until 4.5 million years ago, none was bigger than 43 ft (13 m). At this point there was considerable species diversity in the group, and so gigantism has occurred in several different lineages. The first thing to note is that there is nothing about baleen whale lifestyle that demands huge size, since they had been filter feeding with baleen exclusively for millions of years before their growth spurt.

Today, the baleen whales exploit highly productive waters near the poles. The productivity of particular area of these waters is unpredictable, both in terms of how much food is available in a particular place and when it will be most bountiful. This is because the productivity is a function of the local ocean current patterns, which are not constant, and the frequency and intensity of storm weather in the preceding months, which is even more unpredictable. The flip side of this unpredictability is that there can be areas of food bonanza for the filter feeders that can find them. Large baleen whales are ideally suited to take advantage of such abundance thanks to their large mouths, capacious stomachs, and high metabolism (allowing rapid digestion), allowing them to gorge themselves. Their size also helps them find these food concentrations for a couple of reasons. First, as discussed in Chapter 1, metabolism increases at a slower rate with mass compared to the size of stores that can be carried, so going without food for a while is easier for larger animals than smaller ones. And second, as also discussed in Chapter 1, the cost of moving about, especially by swimming, is no higher for larger animals, but swimming speed is likely to be faster and so they are more able to search the vast oceans effectively for food bonanzas. Our best answer to the question of why baleen whales are so large, therefore, is that changing climate patterns (which affect ocean currents as well as the weather) in the last few million years have led to more unpredictable and concentrated food sources for these filter feeders.

Pygmy right whale
Caperea marginata
Length: 11 ft (3.5 m)
Despite being the smallest of the fifteen species of baleen whale, the pygmy right whale is still bigger than almost every single living land animal.

A Fishy Riddle

Another baffling question is why whales are bigger than fish, especially as there is much greater diversity among fish than baleen whales and fish have been around a lot longer. As discussed earlier (see page 72), three separate lineages of sharks have produced some massive filter feeders, but most of the baleen whales are bigger than these—sometimes spectacularly so. One possible answer to the riddle is that only a combination of high metabolism and large size makes it possible to exploit polar waters. However, while high metabolism is vital to efficient functioning in the cold waters near the poles, this is only part of the story. There are fish in very cold polar waters, but they are not the biggest fish and their body temperature pretty much tracks that of the water around them, making them relatively sluggish. Thus, foraging modes that need constant movement (like large filter feeders) or sudden bursts of acceleration (like predation on substantial prey) need a high body temperature that in cold polar waters can be delivered only by a large-bodied endotherm. So, it might be that the partitioning of the seas, such that whales but not large fish exploit the polar regions, is all about their different metabolic strategies. It could be that temperate and tropical seas don't experience the unpredictability and intense concentrations of food seen in the polar seas today, and so the climate change that triggered gigantism in the baleen whales did not produce the same selection pressure for the sharks.

▲ As in other rorquals, Bryde's whale (*Balaenoptera brydei*) has pleats on its underside. When it opens its mouth to engulf large volumes of water and prey, the pleats expand to increase the size of the mouth even further.

Design Differences

A third question that taxes evolutionary biologists is why whales are bigger than extinct marine reptiles. Marine mammals have been around only since mammals returned to the oceans after the Cretaceous–Paleogene extinction event wiped out the dinosaurs and most other organisms 66 million years ago. Previously, throughout the time of the dinosaurs, several lineages of terrestrial reptiles also evolved to exploit the oceans and eventually cut their ties with the land entirely (analogous to what happened with whales; see page 96). Despite extensive radiations over more than 100 million years, giant filter feeders like the baleen whales didn't flourish among the marine reptiles. This seems odd, especially as filter-feeding fish were present in the oceans at the time. One possibility is that the metabolism of the marine reptiles was dissimilar to that of marine mammals and the polar waters were not as productive at the time, so the reptiles had no advantage over the fish that had already filled giant filter-feeder niches.

▲ For giant whales to fuel their metabolism from tiny krill, they must eat astonishing numbers of these minute crustaceans.

◀ The strength needed for a 25-ton (23-tonne) humpback whale (*Megaptera novaeangliae*) to throw itself out of the water (likely a form of social signaling) is humbling.

Another likely explanation focuses on the mammalian connection between the mouth and the digestive tract, which is a lot more complex than in reptiles. This is why snakes, for example, can swallow comparatively huge prey items whole (see page 173). However, one thing mammals can do is completely seal off the back of the mouth so that it can fill with water without this running down the throat—this is what allows us to use mouthwash, for example. More pertinently, this voluntary control at the back of the mouth is vital to the way the baleen whales feed. This adaptation may therefore have made filter feeding much easier for mammals than reptiles, and so this feeding strategy evolved in one group and not the other. To give you an idea of how a mammal's throat morphology constrains the size of things it can eat, a blue whale—despite its colossal size—can't swallow anything larger than a beach ball, and certainly not you, me, or Jonah!

▲ Humans have complex anatomy at the back of our mouths to control our swallowing. Snakes have a simpler anatomy, allowing them to squeeze huge meals down their throat.

Jaws of Death

The book and film *Jaws* are pretty misleading. Despite its scaremongering, you are very unlikely to be attacked and killed by a shark: there are about eighty recorded unprovoked attacks on humans by sharks a year, of which about 10 percent are fatal. To give you some perspective, each year 500 people in the US are struck by lightning (of which about 10 percent are also killed), while only about fifteen are attacked by sharks.

▲ Even a small great white shark (*Carcharodon carcharias*) would have no difficulty biting clean through a surfboard.

Shark Attacks

Although there are nearly 500 species of shark, almost all attacks on humans can be traced to just four species: the great white (*Carcharodon carcharias*), the tiger shark (*Galeocerdo cuvier*), the bull shark (*Carcharhinus leucas*), and the oceanic whitetip shark (*Carcharhinus longimanus*). There are a couple of reasons for the low fatality rate. First, sharks have a broad diet and will take an interest in anything that might be prey. While they might take a bite from you to see if you are worth further investigation, there is not a lot of good eating in a human compared to a big fish or marine mammal, so the shark will often lose interest. Second, sharks want to avoid being injured themselves by wrestling with large prey at close quarters, so their general tactic is to bite and retreat, repeating this until the prey becomes weak from blood loss. If you are near shore or a boat, or other people nearby can come to your aid, then the shark's initial retreat might give you the necessary respite to get yourself out of the way of further harm.

The Great White

In one way *Jaws* is correct, in that the size of the great white shark can indeed be impressive. Exceptional individuals are more than 20 ft (6 m) long and weigh nearly 4,400 lb (2,000 kg); only the whale shark, basking shark, and manta ray are bigger (see pages 72–75). The great white grows to such a large size for three main reasons. First, the only known predators of large great whites are killer whales, and even these generally prefer easier kills. Second, and more likely, the great white's own feeding preferences have selected for large size. The bigger individuals specialize in preying on marine mammals, and clearly need to be large to hold their own in aggressive encounters with elephant seals, dolphins, and the like. Great whites are often also found scavenging on the corpses of dead whales, and this is considered to be an important source of food for them (it may have been even more important before the era of commercial whaling). Here, large size is essential for the great whites to compete with other sharks for priority access to the corpse. Finally, the scale of this animal is so immense that it retains heat well in its body cavity. From experiments where scientists have fed great whites fish with temperature loggers hidden inside, we know that their stomach can be as much as 27°F (15°C) warmer than the surrounding waters. This raised temperature allows more efficient digestion and enables the shark to take full advantage of a bonanza food source such as a large whale carcass that might be available for several days. Unfortunately, the great white's unusual thermal physiology has not undergone much scrutiny, as it is not practical to keep these giants in captivity—the animals generally refuse to eat, even to the point of starvation.

Having considered the selection pressures that have led to the great white's large size, we should ponder what factors stop them from becoming any bigger.

Great white shark
Carcharodon carcharias
Length: up to 20 ft (6 m)

This great white is in the act of ripping a huge chunk of blubber from a whale carcass.

▶ A seal looks like a tiny snack when seen in the jaws of the huge bulk of a great white shark.

As we saw in Chapter 1, energy availability limits the size of all predators, but for great white sharks in particular, there is also the issue of maneuverability. Many marine mammals can swim fast and are very agile, and having a larger size might make it more difficult for them to catch their smaller, more nimble prey and wouldn't help in dispatching them.

Another thing *Jaws* does get right is that there are plenty of reliable reports of great whites attacking and even sinking boats. Again, this is testament to the size and strength of these marine predators, and also their broad diet and willingness to test just about anything they come across to see if it is potentially edible.

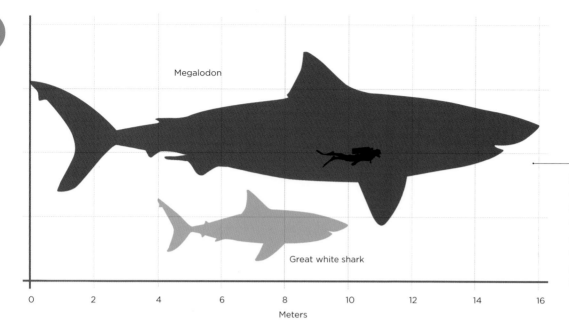

Megalodon

Great white shark

Meters

Megalodon
Carcharocles megalodon
Length: up to 52 ft (16 m)

This marine monster once shared the oceans with ancestors of today's great whites, but it utterly dwarfed them.

Megalodon

From about 23 million years ago to 2.6 million years ago, a predatory shark that dwarfed the great white roamed the world's oceans. Megalodon (*Carcharocles megalodon*) is thought to have reached 52 ft (16 m) in length and weighed anywhere between 33 tons (30 tonnes) and 66 tons (60 tonnes.) There is a lot of uncertainty here because, like all the sharks, the megalodon's skeleton was made of cartilage rather than bone, which decays after death and is much less likely to fossilize. We know the creature was a predator because fossil hunters have found an abundance of its teeth, and these were really robust and sometimes reached 7 in (18 cm) in length. Other than that, all that has been found are a few vertebrae and some coprolites (fossilized feces), which must have come from a giant marine predator and are generally assigned to this species. Scientists assume that a creature this big must have hunted giant prey like the baleen whales, and indeed whale bones with damage most likely inflicted by megalodon teeth have been found. This suggests that the shark had fearsomely powerful jaws and didn't have to attack the soft underbelly of whales—it could have inflicted terrifying bone-crunching damage no matter where it chose to bite. There is also evidence in the stable isotope ratio, or chemical signature, of the teeth that suggests the megalodon was eating large animals.

Why the megalodon died out is unclear, but it was found across all oceans except in the cold waters near the poles, and the timing of its extinction seems to coincide with a period of general cooling of the Earth and expansion of waters that were likely too cold for the species. It had always faced competition from predatory whales (see page 92), but the cooling of the world's oceans might just have given the whales an edge. Killer whales (see page 94) may also have become increasingly sophisticated in their pack-hunting techniques at this time, and this possibly contributed to the megalodon's demise. In addition, the period was one of reduced marine productivity in general, and reduced whale diversity and abundance in particular; if the megalodon was a specialist on giant whales and too slow to maneuver in pursuit of smaller prey, then there might simply have been insufficient food to support a viable population of these giants. The demise of the megalodon may have been down to a combination of all these factors, but the extreme specialization of feeding only on giant whales might explain why it went extinct but the great white (which was around at the same time) survived.

Big Fish Unrelated to Sharks

Megalodon might well have been the biggest fish ever known. Its nearest rival was a Jurassic species in the genus *Leedsichthys*. Although this giant evolved from bony fish, it developed a cartilaginous skeleton, so again the fossil evidence is very limited. However, the current best guess for its length is 52 ft (16 m).

The largest bony fish alive today is the extraordinary ocean sunfish (*Mola mola*). It has a disk-like body with long fins at the top and bottom, and the largest individuals might measure 14 ft (4.2 m) from fin tip to

fin tip and 10.5 ft (3.2 m) from front to back, and weigh as much as 5,000 lb (2,300 kg). Like *Leedsichthys*, it also evolved from bony fish, and again much of its skeleton is cartilage. This suggests that for fish, but not whales, a cartilaginous skeleton is essential for maintaining a huge size. It may be that the higher-energy lifestyle of endothermic whales makes their need to swim constantly (in order to generate upward force to resist the weight of their bones) more sustainable. I am not quite sure this is the reason, however, because the ocean sunfish also shows high activity levels, feeding nearly constantly on jellyfish. It is truly a fish like no other, and you have a reasonable chance of seeing one—in the wild, they are known to bask broadside at the water's surface (perhaps to warm up after diving to the depths to feed), and in captivity they can be found in a few large public aquariums around the world, where the sight of this beast alone is worth any entrance fee.

Ocean sunfish
Mola mola
Weight: up to 2.5 tons
(2.3 tonnes)

One of the very largest fish, and one of the strangest looking, ocean sunfish are known to bask on the surface to help them warm up after deep dives.

▲ We imagine that the megalodon primarily hunted large whales—it would have been powerful enough to catch these marine mammals and small prey would likely often have been too maneuverable for it.

◄ Like all sharks, megalodon regularly shed their teeth, so stunning fossilized remains like this are relatively commonplace.

The Sperm Whale

Unlike the baleen whales, which are generally enormous, the toothed whales are only fairly big and most are dolphins not much bigger than a person. However, a few do reach enormous sizes and one, the sperm whale (*Physeter macrocephalus*), is huge, so we start our exploration of the toothed whales with this species. Although the sperm whale mostly eats smaller prey and is only occasionally a predator of huge prey, it has some very special adaptations to help it in its hunt for food.

In a League of Their Own

The sperm whale is the largest living toothed whale, and although whale diversity has been greater in the past, we don't know of any extinct species that were substantially bigger (see page 92). The biggest male sperm whales reach more than 65 ft (20 m) in length and weigh 63 tons (57 tonnes), and are really in a league of their own—the second-largest toothed whale, Baird's beaked whale (*Berardius bairdii*), is less than 43 ft (13 m) long and 16.5 tons (15 tonnes) in weight. The sperm whale is also the most sexually dimorphic of all the whales, with males often three times as heavy as females. For this reason, males were particularly targeted by commercial whalers. The species as a whole was particularly sought after for the waxy, fatty oil, called spermaceti, contained in a huge organ on the top of its head. When the oil leaked out, whalers thought it looked like seminal fluid—hence the species' common name.

The Spermaceti Organ

The selection pressures that have driven the development of the sperm whale's oil-filled spermaceti organ are not altogether clear. Communication via sound production and prey finding by echolocation are both important to this species, and while the spermaceti organ is certainly involved in the former, plenty of whales communicate and echolocate well without this exaggerated feature. It could be that the spermaceti organ is used to produce extremely loud sounds in order to stun prey. It is true that sperm whales can produce very loud sounds—levels up to 230dB have been recorded (195dB is considered loud enough to rupture your eardrums)—but scientists have failed to stun fish with sounds in this range and so the theory is far from certain.

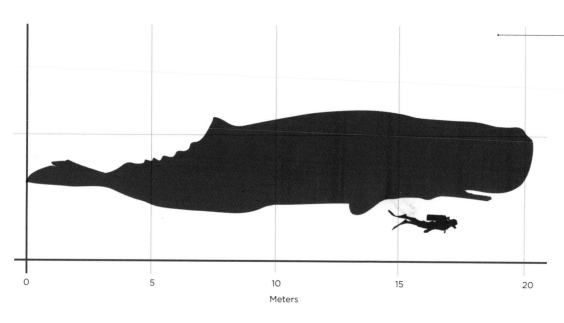

Sperm whale
Physeter macrocephalus
Length: 65 ft (20 m)

The sperm whale is unusually big for a toothed whale, and is generally considered to be the largest living predator.

Meters
0   5   10   15   20

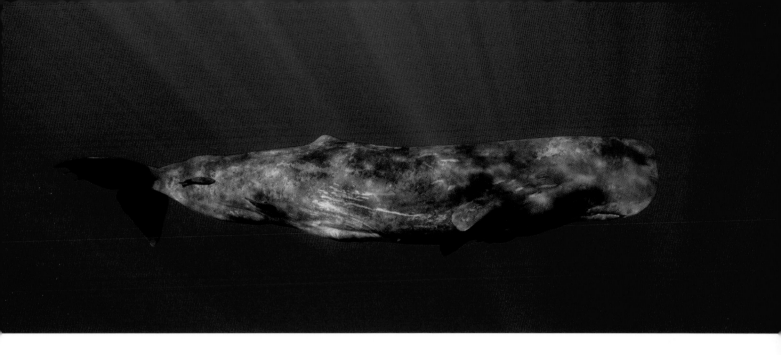

▲ Sperm whales dive to great depths and use echolocation in their search for prey.

▼ The sperm whale's odd shaped head incorporates its huge spermaceti organ, thought to be used in echolocation.

Other theories behind the spermaceti organ include the suggestion that it is a sort of crash helmet to protect the males when they ram one another in competition for females. This behavior between the males has not been observed, but there are well-documented reports of large whaling ships being sunk after been deliberately rammed by sperm whales, most famously the *Essex* in 1820. There is another theory that the spermaceti organ might be useful in buoyancy control. The sperm whale is an extraordinary diver, and can remain submerged at depths greater than 6,500 ft (2,000 m) for periods longer than an hour. It is possible that the spermaceti cools and solidifies early in the dive, and in doing so reduces in volume, making the whale more dense and thus aiding its initial descent. Then, as the sperm whale actively feeds, it warms up from the exercise, which melts the spermaceti, increasing the whale's buoyancy and aiding its return to the surface. I don't find this theory particularly convincing, however, because there doesn't seem to be sufficient blood supply to, and insulation around, this organ to optimize buoyancy control through temperature changes in the spermaceti. There is clearly quite a lot we don't understand about sperm whales, not least the functioning of arguably their most prominent feature.

Baird's beaked whale
Berardius bairdii
Length: 43 ft (13 m)

This is the second largest of the toothed whales, considerably smaller than the sperm whale and with a very different beaked shape to its head.

▲ Despite their large size, sperm whales (*Physeter macrocephalus*) are still vulnerable to predation from killer whales (*Orca orcinus*), but they gain some protection (especially the vulnerable young) from banding together.

Social Behavior

Sperm whales need to communicate with one another because they have complex social lives. Females and young males live together in very stable groups, while larger males depart and lead solitary lives outside the breeding season. Young and injured individuals are vulnerable to attack by killer whales, but sperm whales have been recorded forming a protective circle around a vulnerable individual when the predators appear. They sometimes all have their heads pointing inward, such that the killer whales are faced with powerful tails that could inflict injurious blows (the sperm whale has the largest tail for its body size of any whale), and sometimes with their heads facing outward, in which case the predators might be injured by a head-butt. Sadly, whalers took advantage of this behavior, injuring one individual and leaving it in the water until its companions formed up around it, making them easier for the harpooners to pick off.

Long in the Tooth

Another unusual feature of the sperm whale is its tiny mouth. It has a long, narrow lower jaw, in which are set large teeth, each weighing 2 lb (1 kg). The teeth are well developed but they don't seem essential for predation—sperm whales with few or no teeth, and even deformed jaws, that have been caught still appeared to be well fed. Furthermore, there are reports of whalers opening up the stomachs of sperm whales, to find prey still alive and bearing no teeth marks. The teeth in the whale's upper jaw have reduced to almost nothing through natural selection, and those in the lower jaw fit into sockets in the upper jaw. One theory is that the teeth are used during aggressive bouts between males, and it is certainly the case that mature males often bear scarring consistent with tooth marks of the right size. This brings us back to the theory that sperm whales stun their prey with loud sounds—otherwise, it's a bit of a mystery how they manage to catch their often agile prey at all.

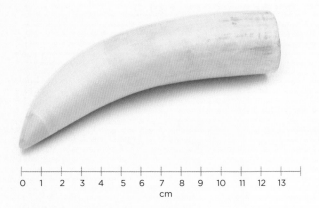

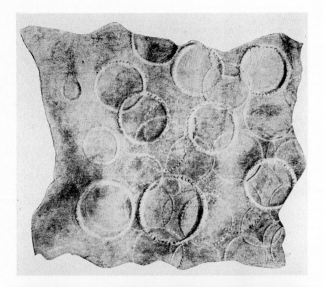

▲ Top: The sperm whale's tooth is long and thin, and quite different from a megalodon tooth (see page 87).

Middle: Scarring on a sperm whale left by the suckers of a squid.

Bottom: A whale leaves a plume of fecal material behind it.

Digestive Disclosures

As discussed in Chapter 7, the longest intestinal worms are found in sperm whales, which themselves have the longest intestines of any animal—more than 980 ft (300 m) long in some individuals. The species has quite a complex digestive system, because it eats larger and more varied prey than the blue whale. The first of the sperm whale's four stomachs is not involved in chemical digestion since it secretes no gastric juices, but it has a very tough muscular lining. This is considered necessary to withstand attack from the claws and suckers of its squid prey, which are swallowed alive and crushed to death in the first stomach prior to chemical digestion in the next three. The hard beaks of the squid remain undigested in the second stomach (around 18,000 were found in one whale) and are periodically coughed up. Although sperm whales predate the very largest squid species in the ocean (as we will also see in Chapter 7), such giants are rare and most of their diet comprises medium-sized squid weighing less than 2 lb (1 kg). In addition, octopuses and fish are regularly found in sperm whale stomachs.

It has been predicted that a sperm whale eats about 3 percent of its body weight a day. It is relatively difficult to calculate the global population size for such a widely dispersed species, but from plausible population estimates it has been estimated that the whales collectively eat six times as much seafood as the total human population. That said, they shouldn't be seen as our competitors, depleting the seas, since they predominantly feed on species that are of no interest to humans as food.

Biologists also argue that sperm whales play an important role in fertilizing the upper levels of the ocean (adding the scarce nitrogen discussed earlier—see box on page 79), because they gather food from great depths but their feces floats until it disintegrates. So much of what we used to know about whales came from sampling dead animals caught in whaling operations. Today, thankfully, this source of information is largely denied to scientists, so those studying whales inevitably take a huge interest in their excrement. Scooping up a whale's poop from the surface of the ocean not only lets you work out what that animal has been eating, but it also reveals its DNA, hormone levels, and various aspects of its health. In all sorts of ways, the sperm whale is a fascinating and still enigmatic organism.

Sperm Whale Ancestors

Between about 15 million and 5 million years ago, various sperm whales living in the prehistoric oceans seem to have been much more suited to hunting larger fish, penguins, and, especially, marine mammals than their modern counterpart. Specifically, they had long, strong teeth firmly set in both jaws, and the jaws themselves were robust. In addition, the jaw hinging and muscle attachments suggest these whales could produce a powerful bite and hold on tight to struggling prey. None of them had the narrow lower jaw of modern-day sperm whales, and instead possessed a long snout in which the jaws were the dominant feature.

Prehistoric Predators

All told, it seems certain that these extinct killer sperm whales were apex predators ready to prey on the biggest animals in the ocean, and were a direct competitor for the megalodon (see page 86). We don't have a clear idea as to why they went extinct, but all the arguments considered for the megalodon apply to these creatures too, except for the issue of avoiding cold waters. It is likely they died out because they were dependent on marine mammals for food, and the diversity and density of these animals declined between 5 million and 3 million years ago. Collectively, these whales are known as the macroraptorial sperm whales, and they include species in the genera *Brygmophyseter*, *Acrophyseter*, *Zygophyseter*, and *Livyatan*. The largest of them all was *Livyatan melvillei*—*Livyatan* from "leviathan" and *melvillei* after Herman Melville, author of the 1851 novel *Moby-Dick*.

Livyatan melvillei
Length: up to 57 ft (17.5 m)

While this giant predatory whale was probably able to attack a megalodon (*Carcharocles megalodon*), other prey would have made easier targets.

▲ Most of the skull bones of *Livyatan melvillei* have been found, so scientists have a good idea of how ferocious its bite might have been. However, more of its other bones need to be discovered before its size can be determined accurately.

To date, only bones from the head of *Livyatan melvillei* have been found, but extrapolating from these, a reasonable estimate for its length is 44–57 ft (13.5–17.5 m), making it about the length of modern sperm whales. It had the largest teeth of any animal (excluding tusks, which are highly modified teeth no longer used for eating), some of which exceed 14 in (36 cm) in length. The skulls of all the macroraptorial sperm whales have a large indentation at the top that is highly suggestive of a spermaceti organ (see page 88). It seems highly unlikely that these leviathans could have stunned giant prey similar in size to themselves, nor would they have needed to with their fearsome jaws and teeth, so this is at odds with the theory for the main function of the spermaceti organ (see page 88). It's amazing that scientists can't figure out the purpose of this feature, but it has certainly been advantageous to a number of sperm whale relatives over the ages.

Most of the macroraptorial sperm whale fossil material discoveries have been made in the last twenty years, so there is real reason to hope that there are more remains yet to be found, and with them we should see a significant increase in our understanding of these ancient giants. One of the selection pressures that may have driven the extinction of these large marine predators was competition with killer whales, with which they likely shared an ecological niche, hunting baleen whales, seals, dolphins, and penguins (see box).

Penguin Prey

Along with marine mammals, it is likely that extinct macroraptorial sperm whales also preyed on prehistoric penguins. The largest penguin living today is the well-known emperor penguin (*Aptenodytes forsteri*), which reaches 4 ft (1.2 m) in height and 100 lb (45 kg) in weight. However, over the last 45 million years numerous other penguin species have been considerably bigger. Nordenskjoeld's giant penguin (*Anthropornis nordenskjoldi*) was not quite as tall as me at 6 ft (1.8 m), but it was heavier, weighing 200 lb (90 kg), while the New Zealand giant penguin (*Pachydyptes ponderosus*) was shorter but probably slightly heavier, sometimes topping 220 lb (100 kg). Last but not least, *Icadyptes salasi* was pretty similar in size but has been found in South America from a point in time when the climate was warmer than now, suggesting that giant penguins could thrive in a variety of different ecological circumstances, not just the extreme cold.

Penguin diversity declined with the rise of the marine mammals, and this reduction in diversity was especially pronounced in the larger-bodied species. This is not surprising, as we know that the emperor penguin faces a huge challenge in rearing its chicks on land. In contrast, mammalian viviparity (where the embryo develops inside the parent) and the fantastic efficiency of feeding on the mother's milk give this group a considerable competitive edge.

▼ Prehistoric penguins were likely prey for marine mammals—just as their descendents are today (perhaps like those in this imaginative artistic depiction).

The Killer Whale

Killer whales are not the biggest of the toothed whales by any means, but at up to 6.5 tons (6 tonnes) they rival the heaviest living land animals in terms of mass. They are also included here because they have key relationships with other large marine species—indeed, no matter which of these animals you consider, there is a reasonable chance that killer whales are a potential predator. In addition, killer whales represent a theme running throughout this book (exemplified by humans and ants): if you band together, you don't have to be the very biggest to have a huge effect.

Killer whale
Orcinus orca
Length: up to 30 ft (10 m)

Killer whales can put on bursts of speed, giving them enough momentum to jump clean out of the water—indeed, they sometimes snatch birds in flight for a meal.

▼ The killer whale's combination of size, power, ferocious teeth, teamwork, and intelligence means that nothing is safe from them in (or even near) the water. As they have a very broad global range, it is also not easy to know when one might appear.

Killer Whale Ecology

Killer whales are ubiquitous in the world's oceans and there are very few places where they haven't been recorded. They are in the same family as the dolphins, but have evolved upward in size and have a very broad diet, ranging from salmon to the biggest of the baleen whales, by way of plenty of animals in between. Killer whales have a complex social system centered around matrilines of related individuals, with the oldest individual being the others' mother, grandmother, or even great-grandmother. An individual can spend all fifty to eighty years of its life in the same social group.

The species is often considered to have the most stable of all animal social groups, which is also thought to be the basis of cultural traditions. For example, individuals of the same matriline call similarly, behave in the same way, and specialize on the same types of food.

Even a single killer whale is a fearsome predator—the largest individuals are almost 30 ft (10 m) long and weigh at least 6.5 tons (6 tonnes). They have sharp yet robust teeth set in powerful jaws. Although prey as diverse as sea gulls snatched from the air and deer taken while swimming between islands have been reported, the main diet of most killer whale pods is either fish or marine mammals—a single killer whale might eat more than 440 lb (200 kg) of fish in a day.

Hunting Techniques

With their global distribution, killer whales eat a wide variety of different fish, but a given pod will generally focus on one particular prey type in a given season. For example, pods off Norway hunt herring, which they catch using a technique called carousel feeding. One or more orcas will circle a shoal of the fish, releasing bubbles to form a curtain around them. They corral the fish into a tighter and tighter ball, which is finally slapped by the whales' tails to stun or injure the fish; these are then picked off at the whales' leisure.

Essentially, all marine mammals are vulnerable to killer whales, but the smaller the species, the more vulnerable it is. There can literally be nowhere to hide in the case of other whales, and seals and sea lions may not even be safe out of the water. Killer whales deliberately ramming an ice floe to dislodge a seal, creating a wave to wash a seal from the ice into the water, and deliberately beaching themselves to snatch animals at the water's edge are all staples of natural history programs. Many of these actions are made more effective by the coordinated behavior between killer whales. The same is true of pods of killer whales harrying really large whales to exhaustion, not unlike the hunting style of packs of African wild dogs (*Lycaon pictus*). They have the advantage over hunting dogs, however, in that by relentlessly preventing a whale from coming up for air, they can soon drive it to exhaustion.

Killer whales are adaptable hunters too. During the whaling era, they soon learned to home in on the noises of human whalers as a signal of the potential availability of food, and attacked injured whales that escaped the whalers or scavenged at the corpses of

▲ Top: Killer whales (*Orcinus orca*) have a particularly tight and complex social system, which is key to much of their success as predators.

▲ Above: Killer whales are also extraordinarily adaptable predators, being willing to hunt prey even out of the water, on beaches or ice floes, or simply flying too low over the surface of the ocean.

dead whales yet to be hauled out of the water. The fish-eating specialists are similarly adaptable—for example, those around Alaska have learned to move along the main lines of longline fisheries, plucking fish from the hooks before fishermen have reeled them in.

There have been very few reported attacks by killer whales on humans in the wild, and none of these has been fatal. This is striking given that there are probably at least 50,000 killer whales around the globe and they are very flexible feeders. Perhaps it is another indication of their intelligence—they rarely misidentify humans as another animal, and those individuals with experience of humans no doubt realize we are not great eating and potentially lots of trouble!

Scaly Sea Monsters

Just as marine mammals are a lineage of once terrestrial mammals that evolved to return to the marine life of their distant ancestors, so a number of terrestrial reptile groups also returned to the oceans. We are familiar today with marine turtles and crocodiles (see Chapter 8), but here we explore some groups of marine reptiles from the age of the dinosaurs that produced some very large individuals but have left no modern descendants.

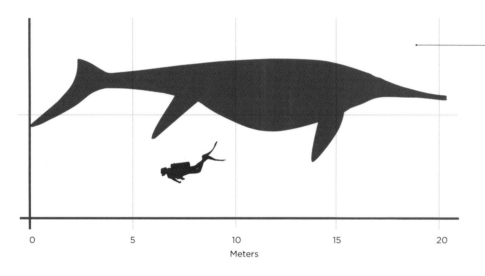

| 0 | 5 | 10 | 15 | 20 |

Meters

Shonisaurus popularis
Length: at least 65 ft (20 m)

One of the biggest ichthyosaurs, *Shonisaurus popularis* is also thought to have had the largest eyes, at around 3 ft (1 m) across, to help it see in deep, dark waters.

▲ *Shonisaurus sikanniensis* was very closely related to *S. popularis* and is also considered to be among the very largest of the ichthyosaurs. The body plan of many ichthyosaurs bears a striking resemblance to that of dolphins and they may also have had similar ecologies.

Ichthyosaurs

Ichthyosaurs (meaning "fish lizards") swam the oceans of the world from 250 million years ago until about 90 million years ago—broadly coincident with the dinosaurs on land. They were very diverse, ranging from extremely sleek dolphin-like animals, to more robust creatures that probably swam using eel-like undulations of the whole body rather than simply a powerful swish of the tail.

From a size perspective, ichthyosaurs could be massive. While some measured just 3 ft (1 m) long, there were certainly several species that were at least 50 ft (15 m) long and others that may well have reached 60 ft (18 m), 65 ft (20 m), or even 80 ft (25 m). The longest individuals are generally placed in the genus *Shonisaurus*, with *S. popularis* and *S. sikanniensis* being the species most commonly cited as the biggest. All ichthyosaurs were carnivorous, but they had a range of different feeding adaptations. The most common of these was a long, thin snout, which could be moved quickly and was filled with narrow, sharp teeth—ideal for catching fast-moving medium-sized fish. Other ichthyosaurs had shorter, more robust jaws containing

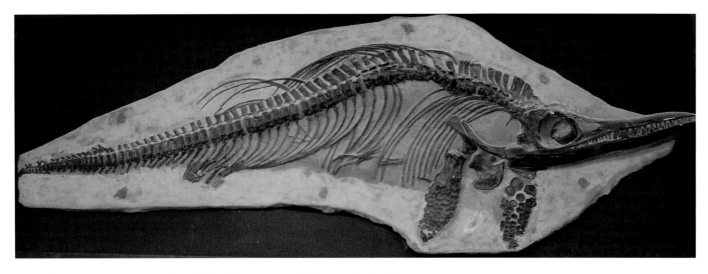

▲ This fossil shows the large number of "finger" bones many ichthyosaurs had, giving their front flippers extreme flexibility to enhance their speed and maneuverability.

sharp but short, resilient teeth that often sheared against one another. This is suggestive of a diet involving larger prey (of similar size to the animal itself), including other marine reptiles or large fish. A small number of species had exceptionally powerful and robust jaws, with flatter, strong teeth; these are very suggestive of a diet that involved cracking the protective shells of mollusks—indeed, ammonites with puncture marks from the teeth of such predators have been found.

The least common ichthyosaur feeding method seems to have been filter feeding on very small prey—this is known from only a handful of species. As already discussed (see page 83), this mode of feeding likely evolved less naturally in reptiles than in either fish or mammals. Since the biggest fish and marine mammals are filter feeders, this might go some way toward explaining why ichthyosaurs did get very large, but not quite as large as the biggest whales. There is evidence that some ichthyosaurs at least might have been endothermic. They are considered to have been open-water cruisers, and it is possible the heat produced by swimming for long periods of time was retained in the body of the animal, through a combination of its large size, minimizing blood flow to the exterior of the body, and a thick layer of insulating fat.

Ichthyosaurs died out before the Cretaceous–Paleogene extinction event 66 million years ago. It seems that there was a reduction in diversity tens of millions of years before their final extinction, and that the very largest forms were among those that disappeared relatively early on. We really don't have a good understanding of what caused their extinction, but it doesn't seem to have been related to competition with other marine reptiles, and my instinct is that it wasn't caused by competition with sharks or other large predatory fish either. If such competition was important, and ichthyosaurs were endothermic, then we might have expected preferential extinction in warmer waters, but there is no sign of that. The marine reptiles that dominated after the ichthyosaurs went extinct were less adapted for open-water cruising and more adapted for ambush predation, relying on fast acceleration and only short bursts of energy. Ichthyosaurs seem designed for efficient cruising at relatively high speeds, but there is no getting away from the fact that constantly being on the move is expensive, and swimming at higher speeds particularly so. It is possible that ecosystems changed such that there was lower productivity generally. In this scenario, an open-water cruiser like an ichthyosaur—which was constantly spending energy in search of food, ate high up in the food chain, was large, and was endothermic—was too disadvantaged energetically to make a living. The fast-moving prey may have evolved to be more efficient swimmers and were just too expensive in energy terms for the ichthyosaurs to chase.

▼ In contrast with terrestrial dinosaur fossils, many complete ichthyosaur and other prehistoric marine reptile skeletons have been found.

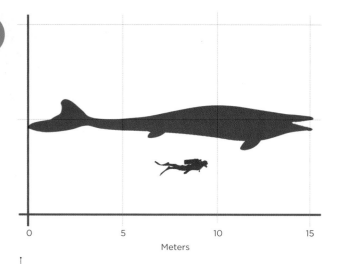

Mosasaurus hoffmannii
Length: 55 ft (17 m)
This species is considered one of the largest of
the mosasaurs, which rose to prominence after
the extinction of the ichthyosaurs.

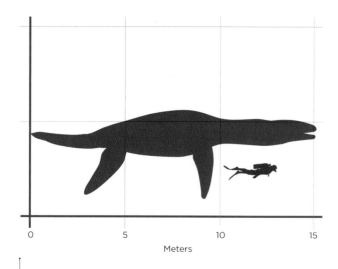

Pliosaurus
Length: up to 50 ft (15 m)
Some of the largest pliosauromorphs were
members of the genus Pliosaurus.

Mosasaurs

After the ichthyosaurs died out, another group of
marine reptiles rose to prominence—the mosasaurs
(meaning "river lizards"). These were also large—
rarely shorter than 13 ft (4 m) and sometimes as long
as 55 ft (17 m). They were similar to ichthyosaurs
in many ways, being apparently endothermic and
ranging from specialists on ammonites and fish, to
top predators that hunted beasts their own size.
However, the body form of these animals doesn't
seem to be designed for open-water cruising, but
rather for lurking and pouncing on passing food.
The endothermy seems surprising—a sit-and-wait
predator can remain immobile longer if it has a low
ectothermic metabolism, and this is especially true
of an air-breathing aquatic organism. My feeling is
that mosasaurs might have been quite active hunters,
but not out on the featureless open oceans like
ichthyosaurs. Instead, they may have favored the more
complex environments of shallower coastal regions
with coral reefs and kelp forests to hide in, and more
turbid waters to mask their approach.

The biggest mosasaurs include *Mosasaurus hoffmannii*,
which has been estimated to reach lengths of 55 ft
(17 m); *Hainosaurus* species, which were anywhere
from 40 ft (12 m) to 56 ft (17 m) in length; and
members of the genus *Tylosaurus*, which are thought
to have attained lengths up to 46 ft (14 m).

Plesiosaurs

The plesiosaurs (meaning "near lizards") were another
major group of aquatic reptiles that returned to the
sea after evolving from terrestrial ancestors, and lived
coincidently with the ichthyosaurs and mosasaurs.
What differentiates them is their very different body
form and thus mode of swimming. While propulsion
was generated in members of the other two groups
by beating the tail or even undulating the whole body,
the plesiosaurs had four huge flippers that they used
to "fly" through the water. However, they were similar
to the other marine reptiles in a number of ways:
they were air breathers (like all reptiles); they never
ventured onto land; they gave birth to live young (very
unusual in lizards, but common in extant sea snakes);
and they were possibly endothermic.

There were two major types of plesiosaurs:
plesiosauromorphs, which had a small head at the
end of a long neck; and pliosauromorphs, which had
a large head on a short neck. The feeding strategy of
the long-necked form is little understood. Originally,
it was thought that they were ambush predators that
could lie in wait and then dart their long neck out to
snatch prey passing in the general vicinity, without
the effort of accelerating their large body. However,
it seems the neck was likely too stiff to allow this.
Instead, my theory is that these animals swam through
turbid coastal or estuarine waters toward their prey,

which wouldn't have seen the small head until it was almost upon it, and couldn't spot the large body trailing behind in the gloom. Another possibility is that plesiosauromorphs swam along the bottom and used their head to plow through the sediment to capture prey buried out of sight. Strange fossilized seabed furrows have been discovered that some scientists connect with exactly this type of long-necked plesiosaur feeding style. However, I wonder whether the head and dentition of these creatures were robust enough to deal with any rocks also buried in the sand. The short-necked plesiosaurs would certainly have been more comfortable cruising around in search of relatively large prey to attack.

Some of the biggest pliosauromorphs are members of the genus *Pliosaurus*. There are no complete skeletons of these, so size estimates are contentious—the largest individuals might have been 50 ft (15 m) long and weighed 50 tons (45 tonnes), although other calculations based on the same material come up with lengths in the range of 33–42 ft (10–13 m). Long-necked individuals in the genus *Elasmosaurus* are estimated to have been 33 ft (10 m) long, including a 23 ft-long (7 m) neck.

▲ It is probable that mosasaurs like this *Tylosaurus* preyed on plesiosaurs—as shown here—and vice versa, driving strong natural selection for large size.

▼ *Styxosaurus snowii* was among the largest of the long-necked plesiosaurs at 40 ft (12 m) in length. It likely used its long neck to dart its head towards fast-moving prey.

Chapter 5
GIANTS OF THE SKIES

When we think about flying vertebrates, the first animals that spring to mind are birds. But we shouldn't forget the bats, or that for tens of millions of years the earliest birds shared the skies with a completely separate group of vertebrates, the pterosaurs. We know that the challenges of taking to the skies limit the size of flying birds, because some flightless species have evolved into the biggest birds that ever existed and include the largest species we see today.

The Largest Modern Flying Birds

Although there are some very large insects (as we will see in Chapter 6), the biggest fliers are vertebrates like birds, with a bony skeleton to make best use of their muscles for the energetic demands of taking to the air. Here, we look at the largest extant flying birds, and draw attention to the fact that these species make extensive use of soaring flight. This is inevitable, given that countering gravity through lift generated by flapping wings becomes harder as size increases. Flapping is essential for take-off in almost all birds, and this ultimately limits how large a bird can be and still take to the air.

The Kori bustard (*Ardeotis kori*) is generally credited as being the heaviest living bird that can fly, with exceptionally large individuals weighing 42 lb (19 kg). But there are several unrelated birds that are not much smaller: the California condor (*Gymnogyps californianus*) can weigh up to 31 lb (14 kg) and the wandering albatross (*Diomedea exulans*) up to 29 lb (13 kg), and exceptionally large individuals of the mute swan (*Cygnus olor*) can reach 31 lb (14 kg). That different lineages of birds have evolved such that their largest members weigh between about 20 lb (9 kg) and 45 lb (20 kg) suggests there is some limit imposed by the demands of flight that restricts the size of flying birds to less than 45 lb. This seems very plausible when observing these birds: they might be majestic in flight, but when you watch them take off and land, they often look a lot less graceful.

Kori bustard
Ardeotis kori
Weight: 42 lb (19 kg)

This species is often described as the largest living flying bird, although the great bustard (*Otis tarda*) is only slightly smaller. In both species, males are considerably bigger than females. You can find kori bustards in southern Africa in areas that are not too densely wooded; not surprisingly, they spend much of their time on the ground.

Flying is Energetically Expensive

Flapping flight takes a lot of effort—the huge breast muscles on your roast duck are clear evidence of this. As discussed in Chapter 1, however, the power an organism gets from its muscles increases less than proportionately with mass. This is a real problem for a flier, because flight involves counteracting the force of gravity, and that does increase in direct proportion to mass. Lift is proportional to the area of the wings (which increases less than proportionately with increasing mass) and flapping frequency (which decreases with increasing mass). Thus, there is a maximum size beyond which birds cannot stay in the air through matching the force of gravity with the lift generated by muscle-powered flapping flight. We see evidence for this in all of the big birds mentioned above, which use soaring flight rather than flapping flight whenever they can.

In soaring flight, birds fly in upward air currents with their wings outstretched, letting that air push against the wings to lift them. Upward air currents are formed in a range of conditions and situations. In sunny regions, air above particularly warm patches of ground (such as dark-colored rock) will itself be warmed by the substrate, and will then rise. In exposed areas, where wind hits a solid surface like a cliff or the face of a big ocean wave, it will be deflected upwards. Large birds are expert at taking advantage of these opportunities for gaining height, which costs them no more in energy terms than the effort of holding their wings rigidly outstretched. However, even the most committed soarers use energy-expensive flapping flight to get to areas of upward-directed air, or to help them gain speed in takeoff or lose speed when landing.

Mute swan
Cygnus olor
Weight: 31 lb (14 kg)

The mute swan is probably the easiest giant flying bird to see, since it breeds in almost fifty countries. The trumpeter swan (*Cygnus buccinator*) is very similar in size and is considered to be the heaviest living bird native to North America.

Top:
Wandering albatross
Diomedea exulans
Weight: 29 lb (13 kg)

Long, thin wings give this bird great soaring efficiency.

Above:
California condor
Gymnogyps californianus
Weight: 31 lb (14 kg)

Vultures have broad wings allowing for tight turns.

Gargantuan Glider

Fossils suggest that, since the time of the dinosaurs, some flying birds were several times the mass of the biggest birds we see today. One of the largest of these seems to have had a wing shape more like that of modern albatrosses than vultures, allowing us to make assumptions about the ecology of this extinct giant. We can also argue that the biggest challenges for such a bird might have been takeoff and landing, and that in both cases a strong wind might have been vital for success.

As Big as a Plane

In 1983, some interesting fossils were uncovered during the excavation of the foundations of a new airport terminal at Charleston, South Carolina. These turned out to be the remains of a bird scientists have called *Pelagornis sandersi*, which lived about 25 million years ago. The astonishing thing about this bird is its size: it weighed somewhere between 50 lb (22 kg) and 90 lb (40 kg), and had a huge wingspan of 21 ft (6.4 m), or about twice the size of the largest wingspan of a living bird and similar to that of a light aircraft! There are, of course, large extant birds that cannot fly (see page 116), but it seems that this giant was a flier rather than a runner, given that it is otherwise hard to explain its very long wings and very small legs. Its slender wings were much like those of modern-day albatrosses, and it is assumed that this giant was also an oceanic soarer.

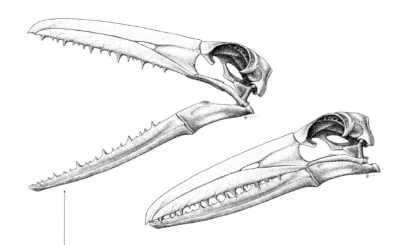

Pelagornis sandersi
Wingspan: 21 ft (6.4 m)

Pelagornis sandersi had unusual extensions to the edge of its bill that look very much like teeth. Closer inspection reveals that they resemble the short, sharp teeth we find in fish-eaters today, like dolphins, so it's probably safe to assume that this ancient bird fed on slippery prey such as fish.

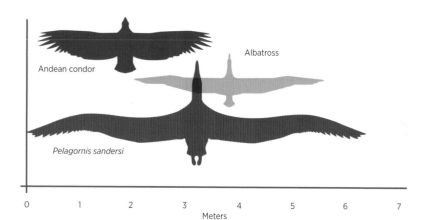

Andean condor

Albatross

Pelagornis sandersi

0   1   2   3   4   5   6   7
Meters

◀ *Pelagornis sandersi* had long, slender (very efficient) wings, much more like those of a modern-day albatross than a (more maneuverable) condor.

The Aerodynamics of Bird Flight

The prime examples of large soaring birds are albatrosses and vultures. Both have a large wing area to increase aerodynamic lift, but they have very different wing shapes. In albatrosses, the wings are very long and thin, whereas in vultures they are much broader.

The albatross wing offers the best compromise between lift and drag (resistance to forward movement through the air), and so leads to very efficient gliding, even in moderate updrafts. However, the cost of this design is reduced maneuverability. Because of their wing shape, albatrosses need a lot of space in which to turn, but this is not a problem in the wide expanses above the oceans, where they fly for long distances in a straight line, benefiting from the winds pushed up by the leading edge of long waves. Greater maneuverability is essential for soaring over land rather than sea.

Vultures often circle within rising columns of air over exposed dark rock that is warmer than the surroundings. These air columns are generally most powerful at the middle, and so the tighter the vulture can circle, the more effectively it can gain height. The broad wings of a vulture allow it to do this, but at a cost of increased drag, so they need stronger updrafts to soar on than do albatrosses, although these updrafts can be more spatially restricted.

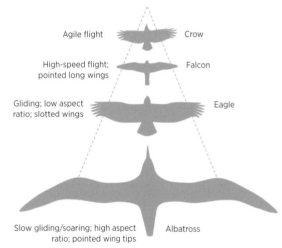

▼ Short, stubby wings (like those of a fighter plane) give maximum maneuverability at a cost of reduced aerodynamic efficiency. Airliners and albatrosses prioritize energy saving by having long, thin wings.

Agile flight — Crow

High-speed flight; pointed long wings — Falcon

Gliding; low aspect ratio; slotted wings — Eagle

Slow gliding/soaring; high aspect ratio; pointed wing tips — Albatross

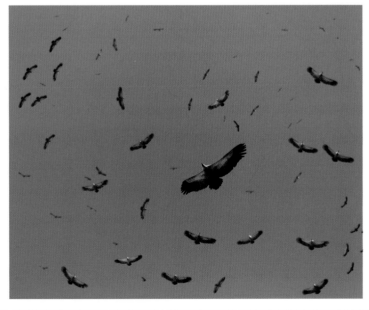

◀ Above left: The black-browed albatross (*Thalassarche melanophris*) is the largest of the albatrosses.

Left: Large aggregations of circling Cape vultures (*Gyps coprotheres*) are becoming less common—as in many other vultures, numbers of the species are dropping alarmingly.

Flying is the Easy Bit!

Gliding might be a good option once a giant bird is in the air, but landing and taking off are a huge challenge for a big animal. Generally speaking, the larger a glider, the faster it flies, and aerodynamic modeling suggests that *Pelagornis sandersi* could not have flown at speeds below 22 mph (36 kph). This is the stall speed, below which its wings would not produce enough lift and it would drop like a stone. This means that the bird would have had to run across the water's surface with its wings outstretched until it was travelling at a minimum of 22 mph (36 kph) before it could take off. This is a challenge: 22 mph is as fast as the very fastest human athletes can sprint. However, it is speed relative to the air that matters for flying, so one solution is to run into a headwind. In a headwind blowing at 11 mph (18 kph), a bird the size of *P. sandersi* would only have to run at 11 mph across the water to take off.

Landing is also problematic for a large flier, since it is generally considered that 11 mph (18 kph) is the fastest speed a bird can be traveling at when it touches down without facing a high risk of crashing and injuring itself. Once again, the answer might be to land into a headwind. But relying on winds for takeoff and landing seems a precarious existence. Imagine if the bird lands to feed on the surface of the sea, only for the wind to drop. It is then stuck there—a sitting duck for any shark or other large marine predator that happens along.

▲ Taking off is hard work for a mute swan (*Cygnus olor*).

▶ This albatross is making use of the drag of the water on its legs to slow it down on landing.

Dining Difficulties

It has been suggested that *Pelagornis sandersi* minimized the need to take off and land by feeding like a frigatebird, simply plucking food off the surface of the sea while still in flight, or harassing other birds into dropping food that it could then grab. Frigatebirds need to settle only when they rest on land and can go weeks at a time on the wing. However, this feeding style requires very accurate flying and it is doubtful whether *P. sandersi* was agile enough—it was more like a supertanker than a speedboat, and needed plenty of time and space to change direction. Modern-day albatrosses land on the ocean surface to feed on slow-moving or floating dead prey, but even the biggest albatrosses can take off by running along the surface of the water until they are going fast enough to get airborne. Thus, it seems likely that *P. sandersi* lived like a giant albatross but would have had to confine itself to parts of the ocean where it could rely on moderate winds.

▲ A frigatebird harasses a red-billed tropicbird (*Phaethon aethereus*) to induce it to regurgitate food.

▼ Frigatebirds spend long periods on the wing, coming to land only to rest and breed. Several species have a distinctive red throat pouch that males inflate spectacularly during courtship.

Latin American Leviathan

The largest known flying bird, a stupendous prehistoric South American species, weighed about as much as an adult human. Looking at its wing shape, we can deduce that it likely had a terrestrial scavenging lifestyle similar to that of modern vultures. When asking why such giant birds are missing from modern ecosystems, we need to consider their specialized habitat requirements.

Argentavis: The Feathered Giant

Fossils belonging to a flying bird that was even heavier than *Pelagornis sandersi* have been found in Argentina. *Argentavis magnificens* lived more than 6 million years ago and is thought to have weighed about 150 lb (70 kg). Again, we can be confident that this bird was a flier, given that it had really long wings as well as air-filled bones for weight reduction. Flightless birds have proportionately very small wings that are mainly used for display and balance purposes, as well as heavier and stronger bones, since weight reduction is not as vital for them as it is for a flier. Further, *A. magnificens* has attachment points on its wing bones for secondary feathers, which have an aerodynamic function and are found only in flying birds. Because of its huge size, this bird must primarily have been a soarer rather than a flapper, but the profound difference in its wing shape compared to that of *P. sandersi* suggests it had a different ecology.

Argentavis magnificens
Weight: 150 lb (70 kg)

Argentavis magnificens seems to have had more of an eagle-like beak than modern vultures, and so likely mixed catching small animals with scavenging on larger carcasses. It probably preyed on animals on the ground from a standing position, like modern herons. However, it could fly, as depicted here.

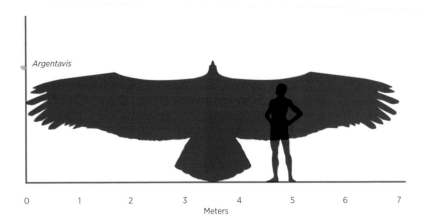

Argentavis

| 0 | 1 | 2 | 3 | 4 | 5 | 6 | 7 |

Meters

Two Different Lifestyles

Remains of *Argentavis magnificens* have been found at several sites within Argentina, but always in rock that was located deep in the interior of the South American continent when the bird was flying, so we can assume it was not an oceanic bird. Paleontologists have yet to discover a complete set of wing bones, but from the sizes of the bones that have been found, they deduce that the bird's wings were not much longer than those of *Pelagornis sandersi*—probably about 23 ft (7 m)—but likely a lot broader. This wing shape strongly suggests soaring over land rather than water (see box on page 105). We assume *A. magnificens* was a scavenger like the modern-day Andean condor (*Vultur gryphus*), feeding on the bodies of large animals that died as a result of accident, disease, old age, or predation.

As in *Pelagornis sandersi*, takeoff and landing would have been real challenges for *Argentavis magnificens*, which aerodynamic modeling suggests probably had a minimum gliding speed of 40 mph (65 kph). It seems inevitable that finding a headwind would have been essential for a safe landing. To gain enough speed for takeoff, the bird would have had alternatives not available to the oceanic *P. sandersi*. First, it could have jumped off a cliff. Initially, it would have fallen like a stone, but if it opened its wings, the speed of the air flowing past them would have offered enough lift once it had fallen 65 ft (20 m) or so to allow the bird to pull out of the dive and glide away. This is essentially how people often take to the air in hang gliders, aided by wind that is deflected upwards by the cliff face. Second, and a little less dramatically, it could have run down a slope to gain sufficient speed—although its leg morphology suggests it was a good walker but not much of a runner. It may be that walking briskly downhill into a headwind was enough to aid takeoff. From this, we can conclude that *A. magnificens* was probably confined to areas where it could be assured moderate to stiff winds.

Why So Big?

It is not exactly clear why *Pelagornis sandersi* and *Argentavis magnificens* evolved to such an enormous size, when being large makes flying more of a challenge. The answer might be found in competition for food, which could have been particularly extreme for *A. magnificens* scavenging large animal carcasses. Such remains would have attracted other birds of prey and mammalian predators, and *A. magnificens* might have had to grow large to hold its own against these in squabbles over meals.

It is possible that no modern-day vultures grow as large because there isn't an environment now that can support them. It is thought that South America 6 million years ago was both warmer and windier than it is now (which would have helped with soaring flight), but still wet enough to sustain plenty of vegetation. In turn, this plant life would have supported a thriving population of animals that ultimately provided food for *Argentavis magnificens*. Quite possibly, there isn't an area of the world today that offers this combination.

For *Pelagornis sandersi*, the same argument about competition might apply if they scavenged on large floating food items like dead marine mammals. A larger size would have allowed larger live prey items to be caught, and large body storage reserves would have seen individuals through long periods when foraging was difficult or impossible. Again, perhaps today there are no areas of the ocean that offer the combination of winds that are reliable but not so strong as to hinder flight, and also enough food for such large birds. Several aspects of a habitat have to be just right to allow really giant flying birds to survive.

◄ Andean condors (*Vultur gryphus*) will willingly share a very large carcass like this dead horse.

Bats: Web-winged Wonders

The other large flying vertebrates we see today are bats, although they never seem to have become as big as birds. Indeed, no fossil remains of bats have yet been discovered to suggest they were ever bigger than the different species of "megabat" now found across Southeast Asia, which rarely weigh much more than 2 lb (1 kg). Bats just don't get anything like as large as the largest flying birds. But why might this be the case?

Lift Limitations

It seems that there have been no giant bats simply because they are not as strong fliers as birds. Bats beat their wings at a slower frequency than birds of the same mass, and they also have less muscle for flapping, so each beat is less forceful. The upshot of this is that bats do not produce as much lift to counteract the force of gravity acting on their mass. This is likely because birds have a more efficient system of breathing (see box opposite). Despite having oversized lungs for a mammal of their size, bats are just not as good as birds at providing the energy needed for flight. As discussed earlier, the demands of flying become proportionately more challenging with increasing size, so bats are forced to stay small.

Yet this doesn't explain why there are no big gliding bats. Bats have never evolved to fill all the feeding niches that birds have, and almost all bats eat insects or fruit. Exploiting these food sources requires agility, and being big doesn't generally coincide with flight accuracy. Although large vultures are more agile than albatrosses, they still need several yards of room in order to turn. You can see this lack of precise maneuverability when vultures come in to land to exploit an animal carcass. They tend to touch down in the general vicinity of the carcass and then walk over to it, because they cannot counteract any gusts of wind quickly enough or select a precise landing spot—they can be confident only of landing in the approximate vicinity.

◄ You might be lucky enough to see a Lyle's flying fox (*Pteropus lylei*) even in urban areas of Southeast Asia.

◀ The Indian flying fox (*Pteropus giganteus*) is one of the world's largest bats, weighing up to 3.5 lb (1.6 kg). As well as the limitations mentioned in the text, another challenge to bats is that mammals give birth to live young. This means that a pregnant bat may have to fly carrying more extra weight for longer than a bird that can lay eggs containing relatively immature young.

Mammal and Bird Lungs

Mammals—including humans and bats—draw air into their lungs and push it out again through the same tube (the trachea). This means that they are not absorbing oxygen half the time because they are contracting their lungs to exhale. It also means that they push some old oxygen-depleted air out of the lungs into the trachea, then inhale it back in again rather than gathering completely fresh air from the outside world. In contrast, birds have a more efficient breathing system that involves a continuous flow of fresh air coming into the lungs. This means that they are better than mammals at gathering the oxygen they need to perform very demanding activities like flying.

Bats synchronize their breathing to their wingbeats. This means that they can save energy by using the same muscle to both push the wings downward and close the diaphragm to breathe out. However, it also means that their wingbeat frequency becomes a compromise between what would be best for the aerodynamics of flight, and what would be best for most efficient breathing.

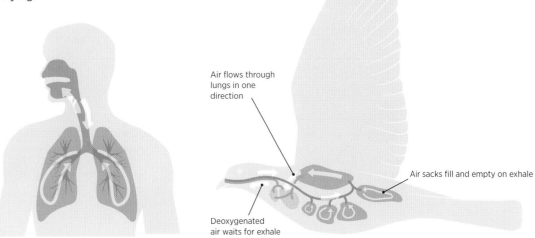

Air flows through lungs in one direction

Air sacks fill and empty on exhale

Deoxygenated air waits for exhale

Titanic Pterosaurs

Pterosaurs are sometimes referred to as flying dinosaurs. This is inaccurate, however, because although they were around at the same time as the dinosaurs (and went extinct at the same time), they are not closely related to them. Ironically, modern-day birds are actually considered by most scientists to be direct descendants of the dinosaurs. As with the prehistoric flying birds, we find evidence of flying giants among the pterosaurs.

Pteranodons: The Giant Pterosaurs

A third group of vertebrates developed flight in addition to the birds and bats. Around the time of the dinosaurs, an unrelated group of reptiles—the pterosaurs—shared the skies with the early birds. Some of these were very large indeed. For one group of pterosaurs, the pteranodons, we have a number of relatively complete fossil skeletons that suggest the largest of them had a wingspan of just less than 23 ft (7 m) and a mass of around 88 lb (40 kg), putting it in the same ballpark as the very largest extinct birds discussed earlier in the chapter. As with fossil birds, the long wings and lightweight bones of these ancient animals strongly suggest they were fliers.

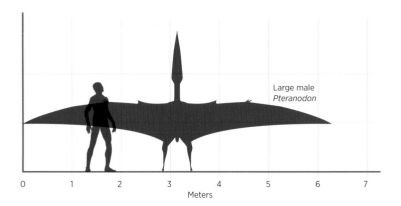

Large male
Pteranodon

Neither Bat Nor Bird

We can be even less sure about how prehistoric pterosaurs might have flown than we are about extinct birds or bats, because there are no living examples we can observe in flight. We can compare pterosaurs to birds and bats, but when we do, we have to be careful to remember that they were very different from both of these vertebrate groups.

Pterosaurs were superficially more like bats than birds because their wings comprised a thin membrane of bare skin rather than skin and feathers, but a pterosaur's wing was quite different from that of a bat. The membrane of a bat's wing is supported by all four of its fingers, which can move independently to vary the shape of the wing. In contrast, the membrane of a pterosaur wing was supported by only a single elongated digit running along the front edge of the wing. This, in turn, made it likely that a pterosaur's wing membrane was thicker, more complex, and more muscular than that of a bat's wing, so that it could be held taut and the wing shape could be controlled. Naturally, this difference means that the aerodynamics of flight of pterosaurs were probably quite different from birds and bats, and in truth they are not currently particularly well understood.

The body of a pterosaur was nothing like that of a bat either. Bats have short hind legs and small heads at the end of a short neck, whereas pterosaurs had long hind legs and large beaked heads at the end of a long neck. This indicates that pterosaurs probably had quite a different lifestyle from bats, which—as discussed previously—almost all eat insects or fruit.

Comparing the morphology of pterosaurs to that of birds, scientists imagine that the large pterosaurs had a lifestyle similar to that of large storks and herons, which walk on the ground and capture small animals like fish and frogs. Concomitant with this, our knowledge of the physics of terrestrial locomotion suggests that pterosaurs were much more comfortable walking on the ground than are bats, simply due to their proportionately longer legs. Fossilized pterosaur footprints suggest they could walk and even run on all four limbs, folding their wings to keep them out of the way.

Pteranodon
Weight: 88 lb (40 kg)

Pteranodons seem to have been marine animals—we find their remains mostly in marine sediments, and we sometimes find bones of fish preserved inside them, indicating that they fed on fish and possibly also marine invertebrates.

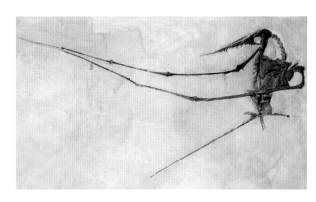

▲ This pterosaur seems to have had an unusually long tail, which can be seen folded over to one side with the wings in this beautifully preserved fossil.

▼ Pterosaurs required very little bone to support their wings. Since bones are heavy, this might partly explain why these reptiles became so big and yet were still able to fly.

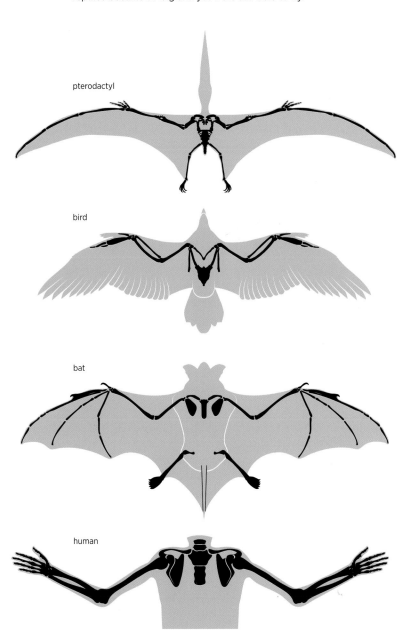

pterodactyl

bird

bat

human

▲ It is thought that pterosaurs could hold their wings out of the way allowing them to walk on all fours, as this artist's impression shows.

Could Pterosaurs Fly?

There seems little doubt that even the largest pteranodons could fly. It is difficult to explain their very large wings otherwise, and fossilized skeletons include many features that we find only in flying animals today, such as hollow, lightweight bones and places for muscle attachment that suggest strong muscles of the type seen only to power flapping flight. That said, if the foraging lifestyle of pterosaurs was similar to that of modern large storks and herons, then they may have spent much of the time on the ground, flying only occasionally to escape predators or to move to a new place to feed. Although herons and storks can walk, sometimes they find it quicker and energetically cheaper to make a short flight to a new foraging ground. Flying also provides the chance for a quick aerial survey to allow selection of a good place to begin searching for food again.

The large size of the pteranodons may have been enough to deter some potential predators, and others could perhaps be outpaced simply by running away. But in response to some terrestrial predators, taking to the air might have provided the best means of escape.

Pterosaur Takeoff and Landing

Given that the largest pteranodons had a similar wingspan to the largest extinct birds but were considerably lighter, the challenges of takeoff and landing (while still considerable) would have been a little less than for the largest birds. There is nothing about the foraging lifestyle of herons and storks that needs particularly precise landing, and we can assume that large pteranodons also tended to forage on estuaries, shallow lakes, marshes, and other predominantly flat, clear, soft-substrate habitats that were relatively kind on bumpy landings.

As far as takeoff is concerned, the most popular current theory is that even the largest pteranodons could have leapt into the air and then flapped their wings for a while to gain height, before looking for some source of rising air. Very small birds can take off from a standing start, but large ones can't and instead need to run with their wings outstretched to take to the air. One theory is that, unlike birds, a pteranodon could have used the muscles that powered all four of its limbs to push against the ground as it leapt into the air, a move that would have allowed it to take off from a standing start. However, it is not clear that this would have been possible. Small birds can take off on the spot because they can simultaneously leap into the air and beat their wings fast enough to generate lift. It seems very unlikely that a big pteranodon could have flapped its 23 ft (7 m) wings quickly. It is more probable that the largest pterosaurs took off from a run, much like large birds today. Their long hind legs would have helped them run fast bipedally, and they would have held their wings outstretched, first for balance and then to give them lift as they ran faster and faster. One advantage for birds (or pterosaurs) foraging on flat, featureless estuaries and marshes is that they can see predators coming from a long way away, and so have plenty of time to take to the air.

Super-sized Pterosaurs

If you browse online, you will find reports of another group of pterosaurs that were not closely related to the pteranodons and that might have been even bigger. If the estimates are correct, these were the biggest natural fliers ever, with wingspans exceeding 30 ft (10 m) and weighing perhaps three times as much as an adult human (570 lb, or 260 kg). These giants have been named *Quetzalcoatlus northropi*, *Hatzegopteryx thambema*, and *Arambourgiania*, but in all cases only a very few bone fragments have been found—indeed, you could fit all the fossil material into a knapsack and still have room left for your lunch. So, our understanding of these animals is necessarily very long on guesswork.

Of these three possible giant pterosaurs, the one for which there is most material is *Quetzalcoatlus*

▲ *Quetzalcoatlus northropi* depicted in flight—the coloration should be seen as artistic license.

northropi, with finds including a single, complete 21 in (54 cm) bone and a few other bone fragments from the same wing. This species lived more than 66 million years ago and might well have been capable of flight; all the pterosaurs we know of were fliers, and the fossil material we have for *Q. northropi* suggests it had the type of lightweight bone typical of the group. However, trying to construct an idea of what a whole extinct animal was like from just one bone requires more imagination than is needed for the other animals discussed here, and that leads to yet more uncertainty.

Although 573 lb (260 kg) is the current best guess for the mass of *Quetzalcoatlus northropi*, in the last ten years researchers on this species have also estimated values as low as 141 lb (64 kg) and as high as 1,190 lb (540 kg), depending on assumptions about the likely morphology of the whole organism. It therefore seems safest to leave speculation on these possible giants until we have more fossil material to guide us. For now, we can be fairly definite that the pteranodons—for which we have more than a thousand discovered specimens—included among their ranks giant fliers at the time of the dinosaurs. Sadly, though, these giants became extinct along with the dinosaurs, when a huge meteor struck the Earth 66 million years ago and the resultant dust cloud shrouded the planet from the sun.

Quetzalcoatlus northropi
Wingspan: 30 ft (10 m)

This pterosaur might not have been too vulnerable to predators on the ground, as its standing height would have made it difficult to sneak up on and its size alone will have deterred many attackers.

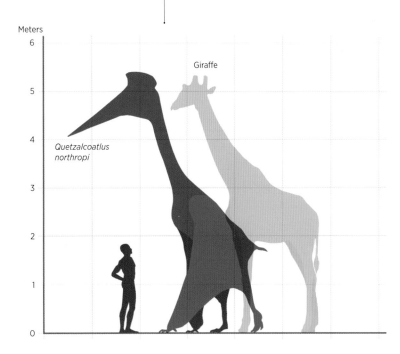

Giant Flightless Birds

The largest of our modern-day birds cannot fly. Not all flightless birds are giants, but the biggest are comfortably taller than adult humans. We know that these birds have evolved from flying ancestors because several aspects of their anatomy—most obviously their wings—can't be explained in any other way. However, the ability to fly is expensive, and for some species the benefits simply stopped outweighing the costs.

The Largest Living Birds

The flightless South American rheas (three species in the order Rheiformes) can reach 90 lb (40 kg), and Australia is home to 110 lb (50 kg) emus (*Dromaius novaehollandiae*) and 130 lb (60 kg) cassowaries (three species in the genus *Casuarius*). The biggest bird of all, however, is the African ostrich (*Struthio camelus*), the largest examples of which weigh 330 lb (150 kg) (see page 120). All these birds evolved from smaller ancestors that could fly, but over time they became gradually bigger, until eventually they were too big to fly. When this happened, it no longer made sense for these birds to have massive wings and the powerful muscles required to beat them. As a result, they now have proportionally small wings that are little used, except occasionally for balance when running and for courtship dancing.

▲ Top: The South American greater rhea (*Rhea americana*) weighs up to 100 lb (45 kg).

Middle: Australia's emu (*Dromaius novaehollandiae*) weighs up to 130 lb (60 kg).

Bottom: Australia's southern cassowary (*Casuarius casuarius*) can sometimes exceed 155 lb (70 kg) in weight.

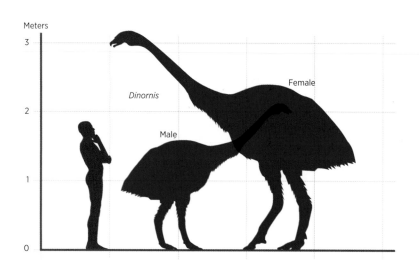

Meters
3

2

1

0

Dinornis

Female

Male

Massive Moa

If one of the modern-day flightless giant birds were extinct and we knew it only from its bones, we would still know it was flightless: its wings would be too short, and the bones in its chest would be neither large enough nor robust enough to support huge flight muscles. There are remains of some extinct giants that show exactly these signs. Until 600–700 years ago, several species of large flightless birds called moa lived in New Zealand, the largest of which—the South Island giant moa (*Dinornis robustus*)—stood about twice as tall as an adult human and weighed 500 lb (230 kg). At the time, New Zealand was heavily forested, so the moa were like modern cassowaries: forest-dwelling omnivores that ate fruit, shoots, seeds, fungi, and any invertebrates and small vertebrates they could catch.

Unfortunately for the moa, the lack of mammals in New Zealand (see box overleaf) made the birds a target for early human colonizers, who arrived around AD 1250. Clearly, a moa would have made a very satisfying meal, as would its eggs. The birds would also have been easy to catch, as they had no ground predators to worry about before people arrived. Tragically, within 150 years of the arrival of humans in New Zealand, all the moa had been wiped out.

South Island giant moa
Dinornis robustus
Weight: 500 lb (230 kg)

There is disagreement over exactly how many species of moa existed, but there is no doubt that some individuals were really big. The shoulder of the biggest of these birds would have been taller than an adult man, without even taking the long neck into account.

◀ Even the largest moa would have been easy prey to humans armed with throwing weapons like spears—and as we managed to wipe them out before they had time to develop any fear of us, they would happily have allowed hunters to get within spear-throwing distance.

An Enormous Eagle

In general, birds of prey are not the stuff of nightmares. For reasons we have already discussed, flying birds tend to be relatively small and lightweight compared to their non-flying cousins. Further, raptors tend to focus on prey that is a good deal smaller than themselves. One reason for this is that they can carry only small animals to the safety of a perch to consume at their leisure, and another is that an injury to their wings sustained in subduing prey could leave them unable to fly. However, the lack of mammalian predators in New Zealand until recently left space for a huge eagle that seems to have preyed on the giant moas.

New Zealand's Top Predator

Moa did have one predator prior to the arrival of humans. Eagles big enough to carry off a man are strictly the stuff of Greek myths, but there was once a giant eagle on the South Island of New Zealand that, according to Māori legend, was large enough to attack children. Haast's eagle (*Harpagornis moorei*) weighed as much as 30 lb (15 kg), but it had notably short wings for a bird that size, spanning "only" 10 ft (3 m). However, its skeleton suggests that it had large, robust bones whose construction implies that they evolved to offer attachment for quite extraordinarily large flight muscles and leg muscles.

Haast's eagle was like a scaled-up version (about 30 percent heavier) of the giant harpy eagle (*Harpia harpyja*) that can be found in the forests and jungles of South America today. The harpy eagle doesn't fly for long periods but instead sits perched as it looks for prey. When it eyes a sloth or monkey, it uses its huge muscles to spring from its perch and take to the air, and then flies very fast and with great maneuverability thanks to its short wings. Such flight is very energy-expensive, but that cost is sustainable if used very sparingly—similar to the sprints of a cheetah (*Acinonyx jubatus*).

Even the largest moa might have been knocked unconscious if struck by the muscly legs of a Haast's eagle weighing the same as a big dog and travelling at 50 mph (80 kph). Once the moa was knocked out, the eagle could land and dispatch it with its ferocious beak, then take its time feeding, secure in the knowledge that no mammalian carnivores would steal its prize. Haast's eagle disappeared around the time of the last moa, suggesting that the flightless birds were indeed its main source of food.

Distant Shores

New Zealand and Australia were once attached to Antarctica, but around 85 million years ago New Zealand split off and began to drift away. At the time, mammals were very minor players as dinosaurs still ruled the Earth. Following the demise of the dinosaurs following the Cretaceous–Paleogene extinction event around 66 million years ago, the early mammals that existed in Australia flourished and evolved into the diverse marsupials we see today. For some reason, however—perhaps just as a result of random chance—early mammals never amounted to much in New Zealand and died out there.

The fact that the only mammals in New Zealand when humans arrived were bats was partly a consequence of this timely separation and the remoteness of the islands. In addition, the prevailing winds and ocean currents around New Zealand generally act to push flying and floating animals and vegetation away rather than drawing them near. It was because of the absence of large predators and competitors that flightless birds could flourish, taking up the niches occupied elsewhere by pigs and deer. New Zealand's unique conditions also meant that it was one of the last places on Earth colonized by people. Polynesian boats like the one shown below were able to withstand long sea journeys enabling Pacific peoples to finally reach New Zealand.

Haast's eagle
Harpagornis moorei
Weight: 30 lb (15 kg)

Although depicted here attacking a large moa, Haast's eagle (*Harpagornis moorei*) likely concentrated on smaller species and juveniles when it could.

Enormous African Avians

Africa is home to the ostrich, the largest living bird on the planet, and also the remains of some of the largest avian species ever known, the aptly named elephant birds, which lived on the island of Madagascar. This survey of giant birds continues by taking a closer look at these living and extinct African avians in an attempt to understand the evolutionary pressures that drove them to evolve such large body sizes.

Ostrich: The Largest Living Bird

As discussed on page 117, the lack of mammalian predators allowed the large flightless moa to thrive in New Zealand. It is therefore reasonable to wonder how ostriches survive today on African savannas they share with the big cats, African hunting dogs (*Lycaon pictus*), and hyenas. In fact, all of these predators do target ostriches from time to time, but catching an ostrich is not that easy. Sneaking up on something almost 10 ft (3 m) tall can be tricky and, given enough warning, an ostrich can stretch its long legs and run at more than 40 mph (64 kph). Its efficient avian respiratory system also means it can maintain this speed for periods longer than mammalian predators can sprint. Even when an ostrich is cornered, dispatching the bird isn't easy: its powerful kicks have been known to kill a human. For these reasons, African predators usually decide that there are easier meals to pursue.

Elephant Birds

Even bigger than the New Zealand moa were the elephant birds from Madagascar. The largest of these (in the genus *Aepyornis*) was more than 10 ft (3 m) tall, weighed anywhere between 770 lb (350 kg) and 1,100 lb (500 kg), and laid eggs weighing about 22 lb (10 kg)—the biggest bird eggs ever, each equivalent to about 150 chicken eggs. Based on morphology and our understanding of the habitat of Madagascar before widespread human influence, these birds probably had a similar ecology to the moa. They probably evolved from a much smaller ancestor that flew to the island, and then became huge and flightless in a habitat that lacked either large-hoofed herbivores or mammalian carnivores. They went extinct sometime around the seventeenth or eighteenth centuries, but had been in decline ever since humans first colonized Madagascar around 2,000 years ago.

Elephant bird

Aepyornis sp.
Weight: 1,100 lb (500 kg)

Despite the giant size of the elephant bird's egg, it would have taken juveniles some time to reach full size. But birds in general have fast growth rates, and year-old ostriches are not much smaller than their parents. Elephant bird juveniles might have had added protection from predators by associating with their parents.

Ostrich

Struthio camelus
Weight: 330 lb (150 kg)

Although the ostrich has lost the ability to fly, it has retained small feathered wings (seen as the feathery mass around its body here) from its smaller flying ancestors. The ostrich uses these in courtship displays, for shading eggs and hatchlings, and for balance when running on uneven terrain.

◄ A beautifully reassembled elephant bird's egg alongside a chicken's egg to offer an idea of scale.

Terror Birds

You'd have to be very unlucky to be killed by any of the giant birds alive today, but things were very different on the South American plains 5 million years ago, when the top predators were the dramatically named terror birds. Here, we look at the current understanding of these birds, before finishing the chapter by considering why giant birds weren't even bigger. Although some were much larger than humans, even the biggest flightless birds have never evolved to the size of the largest mammals and reptiles.

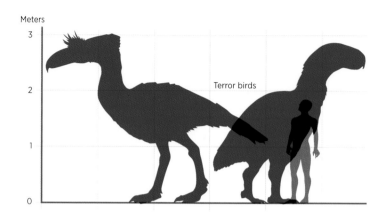

▲ The so-called "Great American interchange" from 2.5 million years ago saw the migration of animals between North and South America, and likely led to the eventual demise of terror birds through competition.

Deadly Predators

The last giants to be considered in this chapter have a name that should capture the attention of those who love the truly scary: the terror birds. These avian predators never killed any humans, but only because there were no humans around to be killed. The terror birds were the top predators across South America from about 62 million years ago until about 2 million years ago. Fossil evidence indicates that the biggest individuals stood 10 ft (3 m) tall and weighed 330 lb (150 kg), and had massive, muscular legs that were capable of propelling them along at speeds of 30 mph (50 kph) and delivering a powerful kick. However, what sets the terror birds apart from all the other flightless birds discussed so far—and marks them out as a definite carnivore—is their powerful, flesh-tearing beak.

A Terrible Demise

Rather prosaically, the terror birds were killed off by geology. Until about 2.5 million years ago, North and South America were separated by an expanse of sea, and quite distinct animal groups evolved on the two continents. In South America, there were next to no large mammalian carnivores and the terror birds (the phorusrhacids to use their scientific name) dominated. However, 2.5 million years ago the sea level fell, leading to the formation of the Isthmus of Panama and creating a land bridge that resulted in the "great American interchange," when animals began to move between the two continents. Bears, dogs, and big cats moved from North America to South America, and are largely considered to have outcompeted the terror birds. One of the largest of the terror birds, *Titanis walleri*, was actually the only member of the family to move north, getting as far as modern-day Florida, where it was never especially common and died out entirely after a couple of hundred thousand years.

Titanis walleri
Weight: 330 lb (150 kg)

It would have been a very brave or very hungry animal that tried to usurp a terror bird from its kill—its powerful legs, sharp claws, and ferocious beak would have been enough to ward off many competitors.

Big, But Not That Big

There are now, and there have been in the past, some really big birds, but not really *giant* birds. The problem of counteracting gravity with wing-generated lift as size increases easily explains why there are no really giant flying birds. However, the biggest living and extinct avians are flightless, and although the birds have been around for tens of millions of years, they have never evolved to anything like the size other vertebrates have achieved repeatedly over a similar timescale. There is no evidence that birds have ever been larger than 1,100 lb (500 kg). To put this in perspective, the lightest dairy cows are about 1,100 lb (500 kg), but elephants can weigh ten times this and some of the largest dinosaurs might have been a hundred times heavier.

The main theory why birds have never reached very large sizes centers on their unique feature—hard-shelled eggs that are almost always incubated through heat provided by one or both parents. The argument is that, as birds get bigger, the eggshell has to become tougher so that the adult can sit on the egg to incubate it without crushing it. But as the eggshell gets thicker, it is harder for oxygen to diffuse through it to allow the chick inside to breathe, and it also becomes increasingly difficult for the chick to break out of the egg when it is ready to hatch.

However, this may not be the whole of the answer. Oxygen diffusion seems relatively easy in birds' eggs exposed to the open air compared with many reptile eggs, which are buried—think of oceanic turtle eggs, buried inches deep in sandy beaches. In addition, it doesn't seem impossible that adult birds could evolve to help crack the eggs when their offspring want to hatch. This behavior has evolved in Nile crocodiles (*Crocodylus niloticus*), which roll eggs around in their mouths when they hear the youngsters calling that they are ready to hatch. One group of birds, the megapodes (family Megapodiidae), have even given up incubation by the parents and returned to placing the eggs in a warm substrate, in the manner of reptiles.

Thus, there might not be anything to stop birds from becoming giants, and that growing to a huge size is just easier for mammals, which give birth to a small number of large live young. For the last 66 million years, it has simply been easier for mammals to fill the niches that require a really giant size—before that, those niches had already been filled by the enormous dinosaurs before the birds came along. Perhaps if the mammals had gone extinct at the same time as the dinosaurs, there might be 5 ton turkeys roaming the Earth today.

Chapter 6

GIANT INSECTS

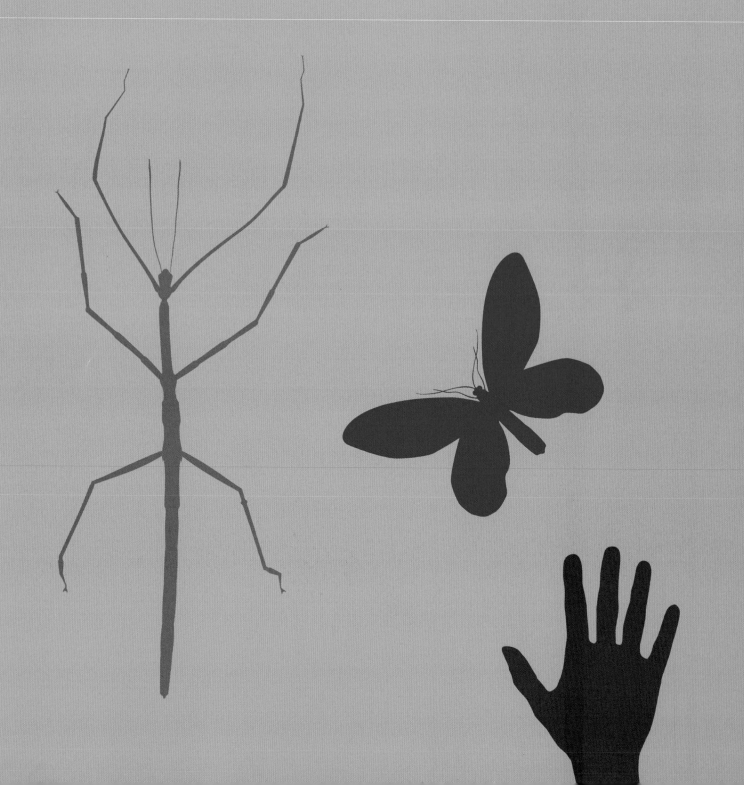

There are more than a million named insect species (compared to only 10,000 birds and 5,000 mammals), and while we don't think of them as being very big, some are pretty impressive. The current record for largest insect is held by a female giant weta (*Deinacrida heteracantha*), found on a small New Zealand island in 2011 by Mark Moffett of the Smithsonian Institution's National Museum of Natural History. This leviathan of the insect world was about three times the weight of a typical mouse or sparrow.

Giant Terrestrial Insects

The giant weta is by no means the only really large insect on Earth, but given the discussion in Chapter 5 about birds, it should come as no surprise that this insect record-holder is flightless—even though flight, at least during the adult stage, is a very common and important feature of insects as a group. Countless species of beetle live in many different environments, and although they have wings as adults, flight is not often commonplace. For this reason, they seem like a good starting point for a search for giant insects.

New Zealand's Terrible Grasshopper

Weta (that's both plural and singular) are thirty-eight species of large grasshopper-like insects (their genus name, *Deinacrida*, is Greek for "terrible grasshopper") native to New Zealand. The country is geographically very isolated, which is why humans colonized it only as recently as 700 years ago (see box on page 118). This isolation meant that, prior to human arrival, there were no mammals on the islands other than bats. The niche of nocturnal omnivores filled by rodents in most parts of the world was filled by weta.

In the large species of weta, the females are much bigger than males, so it's no surprise that the largest individual found was a female. This sexual dimorphism is thought to be a result of competition between males, where smaller individuals are more mobile and so more able to find and mate with females. The record-breaking individual had the further advantage of carrying a full load of eggs when weighed. The giant weta is confined to a single small island measuring only 4 miles (6 km) across. Hence its large size can also be understood as an example of island gigantism, the phenomenon whereby small-bodied animals often evolve larger body sizes when confined to an island (see Chapter 1).

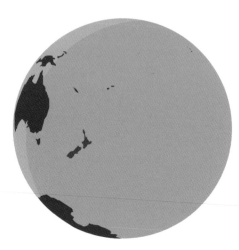

▲ The Pacific Ocean takes up nearly half the globe, and there is a lot of water between New Zealand and the nearest major landmass.

▼ There are many more species of insect in comparison to any other group of organisms: 66 percent of all the species named so far are insects.

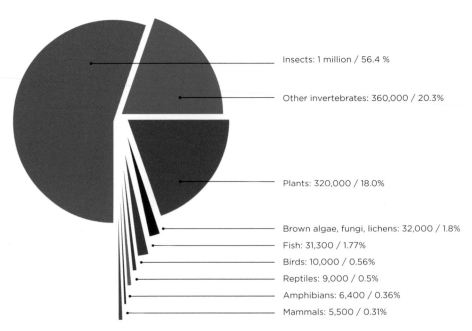

Insects: 1 million / 56.4 %

Other invertebrates: 360,000 / 20.3%

Plants: 320,000 / 18.0%

Brown algae, fungi, lichens: 32,000 / 1.8%

Fish: 31,300 / 1.77%

Birds: 10,000 / 0.56%

Reptiles: 9,000 / 0.5%

Amphibians: 6,400 / 0.36%

Mammals: 5,500 / 0.31%

The Biggest Beetles

Around half of all named insect species are beetles, so it's almost inevitable that this group includes some huge examples. Among their ranks are the aptly named elephant beetle (*Megasoma elephas*), Goliath beetles (five species in the genus *Goliathus*), and the Titan beetle (*Titanus giganteus*), which as adults can be 6–7 in (15–18 cm) long and weigh 1¾ oz (50 g). You can find the Goliaths in the forests of Africa, and the Titans and elephants in South and Central America. Happily, they won't take much interest in you, as they typically eat fruit to keep them going and spend most of the remainder of their time focusing on finding a mate.

The adults of these species are not the true giants, however, weighing only half as much as the immature larvae, which are typical wriggly white grubs that feed in and on rotting wood. These can reach more than 3½ oz (100 g) by the time they've finished growing, at which point they pupate. Within the pupa, the larval tissues are broken down and reorganized to form the very different-looking adult (just as a caterpillar pupates into a butterfly). The emerging adult is considerably lighter that the larva from which it was built. One reason for this is that some of the energy stores of the larva are burnt up in the reorganization process. In addition, some of the mass is devoted to the protective pupal skin within which the transformation occurs, and some larval tissues are unsuitable for conversion and discarded inside the pupal skin. Once the adult emerges, it does not really grow any further—it just eats enough to keep it going in its quest to find a mate and then (in the case of females) somewhere to lay its batch of eggs.

Little Barrier giant weta
Deinacrida heteracantha
Body length: 3 in (75 mm)

The Little Barrier giant weta can weigh 2½ oz (70 g). Although once confined to Hauturu-o-Toi/Little Barrier Island, it has recently been relocated to a number of other very small New Zealand islands where there are no introduced mammalian predators.

◀ Left: Males of the elephant beetle (*Megasoma elephas*) can reach 4¾ in (120 mm) long. This individual took almost three years to reach adulthood and at most might live another year as an adult.

Right: These elephant beetle larvae still have a little growing to do before they molt into their adult form.

Immense Flying Insects

When we think about large flying insects, butterflies and moths are probably the species that come to mind, so these seem like an ideal group to study. However, from the discussion on the previous page about beetles, we would expect that for a given butterfly species, the final mass reached by the caterpillar will actually be heavier than the flying adult that emerges following pupation.

Large Lepidopterans

The giant weta are wingless, and although the giant beetles mentioned above have wings as adults, flying is not a big part of their life. If you seek giant flying insects, look no further than Queen Alexandra's birdwing butterfly (*Ornithoptera alexandrae*), the Atlas moth (*Attacus atlas*), or the white witch moth (*Thysania agrippina*), all of which have wingspans of 11–12 in (28–30 cm). In fact, Queen Alexandra's birdwing butterfly is sufficiently large and slow-flying that early specimens collected by Europeans were downed with the use of a small shotgun.

Because being lightweight makes flight much easier, these big butterflies are nowhere near as heavy as the more terrestrial species mentioned above, and so naturally their caterpillars are not as huge. That said, the cocoons from which Atlas moths emerge were

▲ The wingspan of the Atlas moth (*Attacus atlas*) can reach 12 in (30 cm), but notice how disproportionately short its body is.

traditionally used as coin purses by Taiwanese ladies. Netting was tricky on account of their high-flying habit relative to most butterflies, as they feed high in the rainforest canopy. The weight of caterpillars when they move to the pupal stage is always greater than the flying adult that emerges from the other end of this stage. This is because energy is invested in powering metamorphosis and also because not all of the caterpillar material can be recycled for use in the adult.

Queen Alexandra's birdwing (male)
Ornithoptera alexandrae
Wingspan: 8 in (20 cm)

Females are not quite as colorful as this male but they can be bigger, with a wingspan of 10 in (25 cm) and a body mass of ⅖ oz (12 g).

What's in a Name?

To digress a little, anyone seeing the overall ghostly pale form of a white witch could well understand its name. The Atlas moth's name would similarly come as no surprise to anyone who knows their Greek mythology, but Queen Alexandra's birdwing butterfly takes a bit more unraveling—although "birdwing" clearly indicates its size. Queen Alexandra was married to King Edward VII of England at the time of the European discovery of the species in 1907 by a collector working for Lord Walter Rothschild. The aristocrat came from a well-connected family of English bankers, but he had no interest or talent for banking, and so (perhaps to the rest of the family's relief) devoted himself to studying the natural world.

Rothschild hired collectors to amass a huge private museum of 300,000 bird skins, 2 million birds' eggs, and more than 2 million butterflies, as well as thousands of other specimens. He also had his own zoo, and if you type his name into your favorite search engine, you'll soon find quixotic photos of him on a carriage pulled by zebra or sitting astride a giant tortoise, encouraging it to move forward with lettuce held just out of reach on the end of a stick. The museum was gifted to the public upon his death in 1937 and is open to all in its original site at Tring, just outside London. In a sense, Rothschild's zoo also lives on today, since some edible dormice (*Glis glis*) escaped from his collection and a population of 10,000 or so still thrives in the local area.

Fabulous Flies and Damselflies

The world's largest fly is probably the timber fly *Pantophthalmus bellardi*, which lives in the forests of Central and South America. Its larvae bore into wood and feed within timber until they emerge as adults, which can have a wingspan greater than 3 in (8 cm)—although they seem to be pretty reluctant fliers. They live only a few days as an adult and don't feed, instead focusing entirely on mating and laying their eggs on a suitable tree. You will have to go looking for the fly though, as it will pay neither you nor your picnic any attention.

Another giant fly is the Brazilian Mydas fly *Gauromydas heros*, which has a body length of around 2½ in (6 cm). Little is known about this species, other than the adult is longer-lived than the timber fly and feeds on sweet nectar from flowers—so it may well be attracted to your picnic. If a fly the same size as your peanut butter and jelly sandwich lands on it to feast, remain calm, say "Ah, *heros*" in a knowledgeable voice, and take a photo to impress your friends.

The dobsonfly *Acanthacorydalis fruhstorferi*, from China, can have a wingspan of nearly 9 in (22 cm). Dobsonflies are common across the Americas, Asia, and South Africa, and are often giant—many species have wingspans great than 6 in (15 cm)—but seldom seen. They start life underwater as larvae, much like dragonflies and damselflies, but the adults aren't the agile predators of those groups. Instead, they are clumsy nocturnal flutterers, living for only a few days and generally just mating and laying eggs without showing interest in food.

Megaloprepus caerulatus is the largest of the damselflies and dragonflies (see photo on page 140), with a wingspan of up to 7½ in (19 cm). Like all species in the order Odonata, it is an extraordinarily agile flier, and uses that ability to snatch spiders from their webs to eat. You won't see it hovering over ponds like most of its kind, however, because it lays its eggs high up in the rainforests of Central and South America. It looks for holes in trees that are the right shape and size to collect a miniature pond of rainwater. The attraction of this is that there will be no fish predators in these tiny ponds in the sky to eat the eggs or larvae, but there should be smaller animals in the water for the larvae to hunt. Lots of other insects also use these aerial habitats to escape fish predation, and some frogs even climb trees to find safe places for their tadpoles—although if they choose the wrong pond, they may end up perfect food for a developing damselfly larva.

▶ You have a reasonable chance of seeing the eastern dobsonfly (*Corydalus cornutus*) near fast-flowing water in the eastern part of North America. They are not especially colorful, but can be more than 5½ in (14 cm) long. Note the huge mandibles, which appear to be purely for decoration and intimidation, as the adults don't seem to eat during their few days of existence.

Sting in the Tail

As their common name suggests, tarantula hawks prey on tarantula spiders—although they are actually wasps, not birds. There are at least 250 tarantula hawk species worldwide, in the genera *Pepsis* and *Hemipepsis*, the females of which sting tarantulas to paralyze them, and then drag them off to a burrow they have already dug. Here, they lay a single egg on the spider, which hatches into a larva that then burrows into and feeds on the still living but immobilized spider. After laying the egg, the mother seals the entrance to the burrow to stop any other mothers exploiting the food she has captured for her young, and then goes off to repeat the process for another of her offspring. By keeping the spider alive, the mother wasp ensures the larva's food remains as fresh as possible. The offspring will spend its whole life stage inside the spider, feeding from it, until it finally emerges as an adult, digs itself out of its chamber, and goes off in search of a mate. Male adults lead a much duller life than the females, simply mating and keeping their strength up by sipping nectar from flowers.

Female tarantula hawks are not aggressive towards humans, but if you trap one in your hand it will sting you. According to the Schmidt sting pain index, a measure of the pain caused by Hymenoptera stings, tarantula hawks have one of the most painful stings of any insect. The index sounds very scientific, but is mainly based on the personal experiences of an extraordinary American entomologist, Justin Schmidt, who deliberately subjects himself to the stings of the nastiest insects he can find. Another high-scorer is the bullet ant (*Paraponera clavata*) of Central and South America, so called because its sting is reputed to cause a similar level of pain to being shot!

Honorable Mentions

As discussed at the start of the chapter, there is a bewildering diversity of insects. Unfortunately, owing to space restrictions not every large insect can be included in this book; some of the best of the big also-rans are briefly mentioned here.

Stick insects (order Phasmatodea) include some of the longest-bodied members of the class—some more than 24 in (60 cm) long—although they tend not to be the heaviest since their body is often very narrow (the Goliath stick insect (*Eurycnema goliath*) from eastern Australia is shown below). There are a few giant cockroaches, but the heaviest is probably the giant burrowing cockroach (*Macropanesthia rhinoceros*) from Australia, which can weigh up to 1¼ oz (35 g). Among the big aquatic insects (see also page 138) are the water beetles, which hunt for prey on the water's surface or dive beneath it in search of food. Some of these, including the European species *Dytiscus latissimus*, can be more than 1½ in (40 mm) long and are able to catch small fish.

Also reaching lengths of 1½ in (40 mm) are the worker ants of species like the South American *Dinoponera gigantea*. The list of immense insects could go on and on!

A Huge Hornet

The Asian giant hornet (*Vespa mandarinia*) has a body measuring 1¾ in (45 mm) in length, a wingspan of 3 in (75 mm), and—least appealingly of all—a ¼ in-long (6 mm) stinger. It can be found across tropical Asia, but is most common in rural parts of Japan, where it is responsible for upwards of thirty deaths a year. It takes more than fifty stings to kill a person (unless they are allergic to the venom), something that is most likely to happen when the victim is digging in the ground and inadvertently disturbs a nest, which in late summer may contain up to a hundred workers.

Despite the risk, beekeepers may destroy a nest deliberately, as the hornets prey on a range of large insects, especially honeybee colonies. The native Japanese honeybee (*Apis cerana japonica*) has an extraordinary defense of its own against the hornets. When a hornet discovers a beehive, it starts releasing a chemical pheromone to signal to its kin nearby that there is good food to be had. It takes several minutes of signaling before pheromone levels are high enough for other hornets to detect the chemical, but the bees notice it immediately. They allow the scout hornet to enter the hive, whereupon hundreds of them mob the invader. They surround it in a ball of bees that stops any further signaling pheromone escaping into the air, then they vigorously vibrate their flight muscles. This synchronized action raises the temperature inside the ball, perhaps to more than 113˚F (45˚C), and at the same time uses up the oxygen inside the ball, making it harder for the hornet to breathe. The combination

of intense heat and lack of oxygen kills the hornet, and prevents it summoning reinforcements. More intriguingly, the bees may not always have to go through this expensive defense. It seems that if they signal to the hornet that it has been detected, the hornet thinks better of it and retreats in search of less wary prey.

▼ The Japanese giant hornet (*Vespa mandarinia japonica*) is a subspecies of the world's largest hornet, the Asian giant hornet (*V. mandarinia*), with adults reaching up to 1¾ in (45 mm) in length.

▲ Tarantula hawks are parasitoid wasps, the females of which hunt down and lay their eggs in tarantula spiders. For her own food, the adult female is happy simply to sip nectar from flowers.

Enormous Early Insects

While there are some pretty sizable insects around the world today, they have nothing on the giants that crawled through and flew over prehistoric landscapes 300 million years ago. These included some absolute whoppers, the biggest of which so far discovered were dragonfly-like predators with a wingspan of 28 in (70 cm), or about the same as that of a medium-sized duck. The challenge is to explain why these insect leviathans once existed but no longer do so, and why they were still much smaller than many vertebrate fliers.

Essential Oxygen

The answers to the questions posed above seem to be partly related to the oxygen that all animals need to fuel bodily functions. The challenge is to take oxygen from the surrounding air and deliver it to the cells, and insects have adopted a much simpler engineering approach than large vertebrates like humans. In our case, we actively draw air into our bodies as we breathe in. The internal structure of our lungs is very complex, but is designed to bring our blood into very close contact with the oxygen. This happens in about 700 million tiny sacs called alveoli, the walls of which are perhaps only 0.1 mm thick. Oxygen diffuses across this wall and into the blood, where it is readily absorbed by hemoglobin and then pumped around our body in a complex network of ever-smaller tubes, until it is ultimately delivered close to individual cells, such as muscle cells. There might be 4 in (10 cm) of tissue between our deepest muscles and the outside world, but we can get oxygen from the air to the cells in that muscle in just two or three seconds. The key to this speed is that the oxygen molecules are swept along in the bulk flow of our respiratory and circulatory systems—the only part of the journey they need to make on their own is across the thin walls of our alveoli.

▲ Above left: *Meganeura* was a genus of dragonfly-like insects that lived during the Carboniferous period 300 million years ago and had wingspans of up to 28 in (70 cm).

◀ Left: The similarity between *Meganeura* species and modern-day dragonflies can be appreciated in this perfectly preserved fossil.

▲ Like today's dragonflies, adult *Meganeura* would have been voracious predators, most likely eating mainly other insects but possibly also small vertebrates.

Diffusion Drawbacks

Oxygen moves across the alveoli walls through diffusion. All molecules in a fluid are in constant motion, provided the temperature is greater than absolute zero (−460°F, or −273°C); indeed, the temperature can be thought of as an indicator of how fast they are moving. In the air around us, there are billions of molecules whizzing around, crashing into each other, and then whizzing off again in another direction. Now imagine if you stopped breathing. There would be a higher concentration of oxygen in the air outside your nose compared to the air in your lungs (where, before you stopped breathing, you had been stripping out much of the oxygen). In cases like this where there is a concentration gradient, diffusion acts to even out the variation, so there will be a net movement of oxygen from outside your nose down to your lungs. So, if diffusion delivers oxygen to the lungs, why do we make such a big fuss about breathing? The problem is that it would take a very long time for oxygen to make this journey by diffusion alone.

On average, an oxygen molecule in the air might be travelling at more than 1,000 mph (1,600 kph). The problem is that, on average, each molecule moves less than a micron (a millionth of a meter) before it collides with another molecule, and this collision sends it off in a completely new direction—almost immediately into another collision—and so on. Through this process, the oxygen molecule experiences 6 billion changes of direction each second. Statistically, all these collisions add up to a net movement towards the area with fewer oxygen molecules—in this case the lungs—but it might take half an hour or so for this to result in an equalizing of the concentrations such that the lungs fill with fresh air.

When our bodies are working hard, we need to replace the oxygen in our lungs about once a second, which is why we rely on diffusion for only a very small part of the journey oxygen molecules make from our noses to our cells. This small movement of the oxygen molecules by themselves doesn't slow the process down because the diffusion increases at a rate equivalent to the distance squared. This means that although diffusion is very slow over large distances, it can be very fast over minuscule distances, such as across a cell wall. It might take thirty minutes for a molecule to diffuse across 12 in (30 cm), for example, but only two seconds for it to cross a distance of ⅜ in (1 cm), and a mere 0.05 seconds for it to cross ⅟₃₂ in (1 mm).

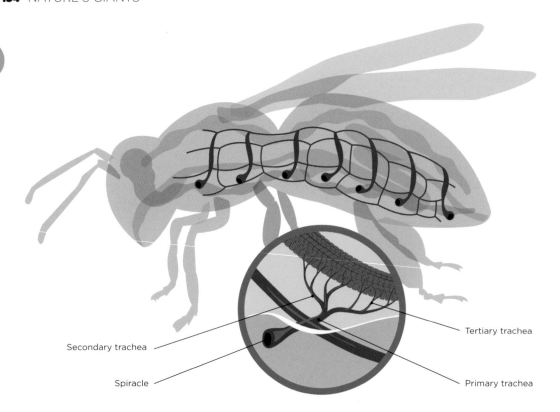

◀ Simplified diagram of the insect respiratory system.

▼ The average and maximum size of insects found in different time periods with different concentrations of oxygen in the air.

Tertiary trachea

Secondary trachea

Spiracle

Primary trachea

Insect Respiratory System

This brings us back to insects, which rely on diffusion a lot more than we do. They have numerous external openings called spiracles, which are positioned at the end of bigger tubes called primary trachea. Each primary trachea splits into several smaller tubes called secondary trachea, which then split into tertiary trachea (see illustration above). This network connects directly to the outside world, with air entering the spiracles and traveling down smaller and smaller pipes to the vicinity of the cells. The key point here is that the insect can pulse its primary trachea to an extent, but most of the process of delivering oxygen from the outside world to cells is through simple diffusion.

The attraction of this design is its wonderful simplicity—an insect's oxygen-delivery system is nothing like as complex as ours, and thus nowhere near as costly to maintain or as vulnerable to breakdown. Yes, diffusion is slow over long distances,

but if insects are small then this is not a problem as the oxygen needs to move only a millimeter or two. However, following the rule that diffusion time increases at the rate of distance squared, an insect that is doubled in size would take four times as long to deliver oxygen to its cells, and one ten times as big would take 100 times longer. This reliance on diffusion is likely to be an important reason why we have never seen (and never will see) insects as big as a human. Depending on your viewpoint, that might be a profound disappointment or it might help you sleep better at night!

We should not see the insects' method of oxygen delivery as inferior to ours in any way. In fact, over the small distances involved, their system works really well—so much so, in fact, that the flight muscles of some insects have the highest rate of energy delivery (and so the greatest consumption of oxygen) of any animal tissue known. The only serious limitation of the

Cenozoic
21% O$_2$

Early Cretaceous
15% O$_2$

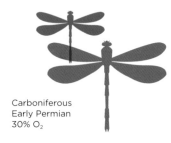

Carboniferous
Early Permian
30% O$_2$

Mid-Carboniferous
21% O$_2$

insects' approach is that it begins to struggle as bodies increase in size. For most animals with a circulatory system, including humans, this arrangement takes up a modest and constant fraction of our bodies. In insects, however, a larger size means having to devote more and more of their body to their oxygen-delivery pipework. This is because the insect copes with the longer time taken by oxygen to diffuse throughout its body by allowing more oxygen to make that journey at any given time.

One way an insect can cope with the problem is to try to have as many of its oxygen-hungry tissues in its thorax as possible, to minimize the distances over which oxygen must diffuse, but the brain and leg muscles are necessarily farther away. In big insects, the pinch points of their body—at the neck and tops of the legs—are increasingly filled with trachea as their size increases. It is likely that these points are the final limitation of insect body size at a given oxygen level in the air. After a certain critical point, it is just not possible to squeeze any more tubes through these constricted apertures to compensate for the increased distances over which oxygen must travel.

Oxygen Availability

The frustrating thing for insects is that most of the air they transport through their respiratory system is useless to them. Currently, the composition of the air around us is about 21 percent oxygen, the rest being mostly nitrogen. Our blood carries only the oxygen, leaving the nitrogen behind in our lungs; in contrast, insects must carry the nitrogen all the way through their pipework along with the oxygen. However, for some periods in the past, oxygen has been more abundant in the air. This was good news for insects: if a greater fraction of the air is useful oxygen, then less air needs to diffuse through their system. This means that a smaller network of trachea will be required for insects of a given size, and the constraint on maximum insect size is relaxed a bit. This is key to why there were giant insects in the past.

In 2012, two scientists from the University of California, Santa Cruz—Matthew Clapham and Jered Karr—assembled a time series of the maximum insect sizes seen since the rise of the insects to global dominance 350 million years ago (see graph below). For the first 200 million years of the series, the maximum size found in the fossil record closely matches the oxygen content of the air, despite considerable fluctuation in the latter. Initially over this period, oxygen levels rose from about current levels to 35 percent, then fell back to current levels, before showing a less spectacular peak of about 25 percent and then declining again. Insect size closely tracks oxygen availability over this period, with the largest insects ever corresponding to the main oxygen peak, which lasted from the Late Carboniferous to the Late Permian, 320–240 million years ago. This close linkage suggests that enhanced oxygen was key to freeing insects up to achieve large sizes: the more oxygen, the bigger the insects.

▼ Change in both the maximum size of insects known and the estimated oxygen concentration in the atmosphere through time.

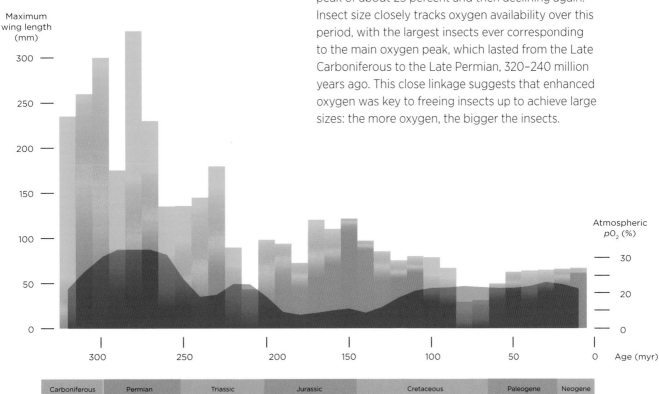

Other Size-limiting Factors

Oxygen isn't the only thing insects require. Just like any other animal, they also need to find enough food to eat and they need to avoid becoming food themselves. Over the past 150 million years, these factors have become more important than oxygen levels in limiting maximum insect size, but could other conditions yield even bigger insects than we see alive today or in the fossil records?

Rise of the Birds

Over the last 150 million years, there has been a change in the patterns revealed in the data collected by Matthew Clapham and Jered Karr (see page 135). From around that point (the start of the Cretaceous period), oxygen availability in the air rose steadily for 50 million years from a low of around 15 percent, stabilizing somewhere near its current value and remaining there for the last 100 million years. But this increase in oxygen is the first instance on the time series *not* linked to an increase in maximum insect size. Indeed, the size of the largest insects actually fell during this period. The likely explanation for this is the rise of the birds 150 million years ago, and their steady adoption of adaptations enabling more agile flight. It seems probable that birds were increasingly able to outcompete or prey upon the largest of the insects, and this problem became increasingly acute for the insects as birds developed improved aerobatic skills. Today, birds are important predators of both flying and non-flying insects.

The birds had two clear advantages over the largest insects. First, aerial maneuverability requires huge power investment to make rapid changes in direction. Being warm-blooded, birds have a high (endothermic) metabolism, and this is associated with much greater maximum power outputs than those achieved by similar-sized low-metabolism (ectothermic) animals like insects. Second, the feathers of birds allow spectacular wing flexibility, whereas insects have much stiffer, less flexible wings.

Sixty-five million years ago, the dinosaurs were wiped out. All sorts of other animal groups suffered in this

▲ This beautiful bird may be called a white-breasted kingfisher (*Halcyon smyrnensis*), but it is about to feast on a large insect.

cataclysm, and it's no surprise that Clapham and Karr's data show a dip in the maximum size of insects after the event. Following this, maximum insect size bounces back, but not by much. Some of the birds survived the extinction event, eventually diversifying into the wonderful acrobats we see today. Worse still for the big insects, mammals thrived in the niches left by the departed dinosaurs, and some of them took to the air as bats. Today, about a quarter of all mammal species are bats. Like birds, they are endothermic, giving them an advantage in acrobatic flight; and they also have a wonderfully flexible wing supported across several fingers—for comparison, just look at what we can do with the flexibility of our own fingers. Thus, for most of their 400 million years on Earth insects have been restricted in size by oxygen availability, but since the rise of the birds and bats they have been further constrained by competition and predation.

◀ A praying mantis having just undergone a molt—the remains of the old exoskeleton once the insect has emerged from it are called exuviae.

▼ On emerging from its chrysalis, a butterfly needs some time to allow its wings to expand before it can fly away.

Alien Monsters?

As a last thought on the subject, imagine we found a planet with really high oxygen levels—say 80 percent—and no birds or bats to spoil the fun. In that scenario, could truly huge insects evolve? Setting aside concerns that such an atmosphere might not be very conducive to life, allowing fires to rage constantly across the planet, bigger insects might be a possibility, but perhaps not colossal ones.

There are two main limiting constraints remaining. First, remember that diffusion rapidly becomes less efficient as distances increase (see page 134). Even with more oxygen in the air, diffusion time still increases dramatically with distance—by a factor of 900, from two seconds to thirty minutes, when distance increases thirtyfold from ⅜ in (1 cm) to 12 in (30 cm). Second, insects are arthropods (like spiders and lobsters, for example), and so have their skeleton on the outside of their body. This exoskeleton must be rigid and strong enough to protect and support the animal inside. Thus, as an individual arthropod grows it must periodically molt its old exoskeleton, and it must stay still while its new exoskeleton forms and hardens. During that time, it is very vulnerable to predation.

If we look across the arthropods as a whole, we find that the bigger-bodied species have thicker exoskeletons. This is not surprising: the exoskeleton must support the organism against the force of gravity. This force will increase with the mass of the organism, and as size increases, mass increases faster than surface area (see Chapter 1). The exoskeleton must fit snugly, matching the surface area of the organism, so as the weight of the organism increases, the exoskeleton has to grow thicker to support it. However, a thicker exoskeleton takes longer to harden, so larger-bodied arthropods have to endure longer periods when they cannot feed and are vulnerable to predators. This is why we find the largest arthropods (the lobsters and crabs) in the sea, where much of their weight is supported by buoyancy and the exoskeleton does not need to be as thick as if the animal were on land. But of all the insects, only a handful spend their whole life in water, since this precludes the flight that brings them so many advantages. Hence, if we are invaded by lifeforms from another planet, we can take comfort that they won't be giant insects.

Aquatic Insects

As discussed in Chapter 4, the real giants of the animal world today are found in our oceans, where much of their weight can be maintained by buoyancy. But why aren't insects more strongly associated with aquatic habitats? Here, we explore why there are so few insects in the oceans and why those associated with freshwater habitats don't grow bigger.

Missing from the Seas

There are more than a million living insect species, but only 1,400 of them live in marine habitats and the overwhelming majority of these mostly stay out of the water and live at its edge. Only 46 species—a group called the gerrid skaters—inhabit the surface of the seas, and most of these live in coastal habitats and river estuaries—just five are fully oceanic. So, there really is a mystery here that requires an explanation. Why, when you see insects almost everywhere you look on land, are they absent from the oceans, which cover 70 percent of the Earth's surface? No one knows the answer to this riddle for sure, but my hypothesis is that it's hard to find somewhere dry in the ocean. Let me explain.

Around 45,000 insect species live in freshwater habitats. Generally, the juvenile stages live underwater, then when the time comes to metamorphose into a winged adult stage, the juveniles use semi-submerged vegetation or anything else projecting above the surface to help them break through the surface tension of the water. They then climb completely out of the water to allow their wings to unfold and dry out, before flying away as adults. On a lake on a really still night, some insects manage to do this while balancing on the floating exoskeleton they have just shed, rather than on vegetation or some other dry surface. On the sea, however, there are precious few opportunities to climb out of the water and find somewhere where newly unfolded wings can dry out.

The marine gerrid skaters lay their eggs on pieces of flotsam—anything that has been washed out to sea or thrown off a boat that floats. Objects that are large enough and buoyant enough to project sufficiently far out of the water that an insect could dry its wings free from the spray of breaking waves, such as a large empty barrel or sizable tree, are very hard to find. This lack of a solid, dry substrate may explain why

▲ Adult mayflies emerge en masse, providing a feast for the local population of predators—and making it very easy to find a mate.

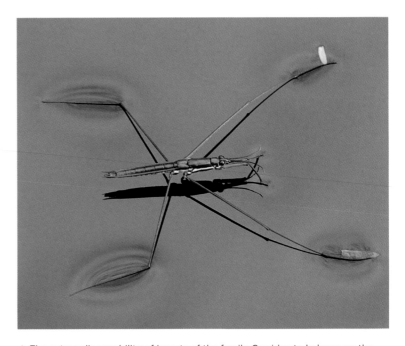

▲ The extraordinary ability of insects of the family Gerridae to balance on the surface of water has led to a plethora of common names, from water skeeters and pond skaters to Jesus bugs.

the oceanic skaters have given up flight. The other 41 skaters that live in salty waters are all associated with habitats where mangroves, emergent plants, or rocks provide such substrates, and all these species lay their eggs clear of the water.

So, exploiting the seas seems to require insects to give up the ability to fly. Flying brings insects key advantages in being able to flee from predators and to search wide areas efficiently in search of food, mates, and other resources. As their name suggests, the gerrid skaters partially compensate for their lack of flight by having very agile movement across the ocean surface, but the lack of other marine insects indicates that having nowhere to dry wings makes the oceans a hugely unattractive prospect.

An Absence of Aquatic Ants

Ants have a huge impact on terrestrial ecosystems—it is estimated that they make up about a fifth of the weight of all land animals on Earth, which is more than all the vertebrates combined, humans included. Given that we have named more than 12,000 ant

species and that they are found all over, it is a bit of a puzzle why none of them makes a living finding food in water. My theory is that the secret to the success of these creatures is their ability to build a predator-safe, long-lasting, expandable nest, and that nest-building underwater is really difficult. Burrowing is much trickier underwater because substrates are soft, and tunnels need constant attention to prevent collapse. Worse still, waterlogged sediments are more mobile at larger scales and the risk of a nest being swept away or buried under mountains of mud after a flood or storm would be quite high.

Making a nest on land and then foraging underwater doesn't seem like a feasible option either. In this case, the small size of ants would work against them—surface tension has more of an effect on smaller organisms, which is why some insects can walk on water but humans can't. It would take too much effort for an ant to break through the water surface every time it left or returned to the water, and as we have already seen, insects can't grow to very large sizes. Hence, while ants have been able to adapt to all sorts of environments on the land, the physics of water has defeated them.

▼ Ants generally try to avoid getting wet, even if they have to cross bodies of water.

◀ The giant helicopter damselfly (*Megaloprepus caerulatus*) can have a wingspan of up to 7½ in (19 cm), the largest of any living member of the order Odonata. Its common name comes from the wings' resemblance to spinning helicopter blades when the insect is in flight.

Freshwater Insects

Returning to the 45,000 insect species that are associated with rivers and lakes, what are the limits to their size? The order Odonata, comprising the dragonflies and damselflies, includes species that are semi- and fully aquatic during their juvenile stage. The largest species, *Megaloprepus caerulatus*, has a wingspan of 7½ in (19 cm) (see page 129), while some ancestors of our present-day dragonflies and damselflies (such as *Meganeura monyi* and *Meganeuropsis permiana*) had wingspans of 12–13 in (30–33 cm). These animals have such slight bodies and narrow, thin wings that they are not particularly impressive in terms of weight, but their wingspans aren't much smaller than those of the very largest butterflies discussed earlier. Dragonfly larvae are

carnivorous, feeding on pretty much anything that comes near them—including amphibian tadpoles and even small fish—and they crawl out of the water to metamorphose into the winged adult form. The aquatic environment where these individuals spend the overwhelming majority of their lives, along with their high-protein diet, should provide them with a good chance to get really huge, but there are a couple of factors that prevent the largest dragonflies from becoming as big as the largest butterflies.

For a start, adult dragonflies are carnivores, catching insects, spiders, and the like while on the wing. This requires them to be exceptionally agile and quick fliers, and agility declines with increasing size. If dragonflies were to grow much bigger, they just wouldn't be able to change direction quickly enough to catch fast-moving prey. The second constraining factor is that metamorphosis from juvenile to adult in the butterflies occurs via a pupal stage, whereas this stage is skipped in the dragonflies, which simply molt from juveniles into adults. The pupal stage takes more time but allows a complete reorganization of the organism into an adult, which has an utterly different body plan from the juvenile. In the case of dragonflies, there must be more of an element of compromise in the design of either the adult or the juvenile body plan, so that the animal can step from one to the other without the total rebuild seen in butterflies. This may mean that dragonfly adults are not ideally shaped for growing much bigger than they already do. Although they may not be the very biggest insects we have, the often stunning colors of dragonflies make them among the most beautiful.

Builders and Artists

We are used to seeing the complex and sometimes giant structures built by terrestrial insects, including termite mounds, anthills, and wasps' nests, but often overlook the creations made by underwater insects, especially caddis flies. The 10,000-plus species of caddis fly inhabit fresh waters all around the world and most build as larvae, in ways that extend their effective size. Some build cases that they live in and carry around with them, constructing them from materials they find around them and binding these together with silk they secrete. Different species have different preferences in terms of the materials they use, from grains of sand, larger fragments of rock or mollusk shells, bark, and seeds, through twigs, to parts of leaves they cut to size and assemble like bricks in a wall. In general, however, they are flexible and will find some way to use whatever material they find around them. As the caddisfly larvae grow bigger, they build extensions to the end of their case.

Some caddisfly cases are strong enough to offer physical protection from a wide range of would-be predators, whereas others instead rely on being big and awkward enough to put predators off. Many fish and frogs are gape-limited predators, meaning that they swallow their prey whole and the maximum size of food item they can consume is limited by the gape of their mouth. Caddisfly cases (especially ones with big twigs sticking out of them) are ideal for foiling these predators.

The French artist Hubert Duprat collects caddisfly larvae and keeps them in tanks, where he carefully controls the materials available for case building. In this way, he and the insects can be thought of as collaborating in the construction of works of art and jewelry, with the larvae weaving precious and semiprecious items like grains of gold into their cases. The cases can be removed without harming the caddis larva, which then immediately starts work on another case.

Other species of caddisfly expand their effective size in another way, by building huge (in comparison to their size) fishing nets. They hang these webs out in running water so that food particles accumulate on them—just like a spider's web. This is a clever way of extending the effective size of the organism in terms of how much of the stream it can cover for gathering food. And again like a spider's web, these structures can be extraordinarily beautiful, even without the intervention of human artists.

▲ Jeweled works created through collaborations between artist Hubert Duprat and caddisfly larvae.

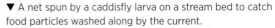

▼ A net spun by a caddisfly larva on a stream bed to catch food particles washed along by the current.

Chapter 7
IMMENSE INVERTEBRATES

In Chapter 6 we focused on just one group of invertebrate animals, the insects. These creatures deserve a section to themselves, since no matter where you are in the world, insects are likely very familiar to you—as the English evolutionary biologist J. B. S. Haldane once quipped, "God has an inordinate fondness for beetles." However, there are many other invertebrate groups, and while we can't cover them all, we will touch on a few interesting giants. We start with the stuff of nightmares: giant spiders.

Super-sized Spiders

Many people suffer from arachnophobia, or an extreme fear of spiders—indeed, studies in Sweden (by no means a hotspot for dangerous spiders) estimate that around 4 percent of adults endure this disorder. So, to put the mind of any arachnophobe at rest, we should say straight up that there are no truly huge spiders on the scale of those seen in horror movies.

The Largest Spiders

The biggest spider by weight is the Goliath birdeater (*Theraphosa blondi*), the largest examples of which can weigh 6 oz (170 g), or not much more than a large hamster. Its common name is a little sensational too, and originated from an early observation of an individual eating a hummingbird. This is not as impressive as it sounds, as birds are rarely part of the spider's diet (it mostly eats worms), and hummingbirds are among the smallest of all the birds, so the bird in this case was almost certainly a lot smaller than the spider. The species could therefore more accurately be named the hamster-sized wormeater—definitely much less the stuff of nightmares. You are unlikely to stumble on one either, as these arachnids live in upland rainforest in South America, spend most of their time in deep burrows, are rarely reported eating vertebrates, and the only mammals preyed on are tiny bats. However, very small birds have occasionally been found in spiderwebs, some South American spiders prey on tiny frogs, and there is even a report of a spider catching a goldfish from a garden pond in Sydney, Australia.

Goliath birdeater
Theraphosa blondi
Weight: 6 oz (170 g)

The size of this spider makes it worthwhile eating for humans, and as such it is actively sought out by local people in the Amazonian rainforest.

Giant huntsman spider
Heteropoda maxima
Leg span: 12 in (30 cm)

Although it is a cave-dweller, the giant huntsman has good vision, suggesting that it lurks in wait for prey near cave entrances.

The other arachnid that lays claim to being the biggest—this time on account of its 12 in (30 cm) leg span—is the giant huntsman spider (*Heteropoda maxima*). You are even less likely to stumble upon this than the Goliath, as it lives in caves in Laos and is rare enough that it was not reported in the scientific literature until 2001.

The largest fossil spider, *Mongolarachne jurassica*, isn't much bigger than the modern-day record-holders. Invertebrates don't fossilize well for the simple reason that they don't have bones. Teeth and bones are hard structures that do fossilize well, and most extinct animals are known to us only from these remains—it is only in exceptional circumstances that softer tissues are preserved. Therefore, there may well have been larger spider species that we don't know about, especially given that they have been around for 300 million years and can make a living in a huge diversity of habitats. However, they probably weren't that much bigger owing to factors similar to those that limit insect size (see page 134). Spiders have a slightly more sophisticated breathing apparatus than insects, but not by much. In addition, they are arthropods, with an exoskeleton that needs to be shed continually as they grow bigger—the cost of repeatedly producing a new protective covering does not make large size a very attractive option in this group of animals.

Whopping Web Spinners

The very large spiders don't build webs. Instead, they are ambush predators that lie in wait for prey to pass, at which point they pounce. A web gives a spider a much wider capture radius than pouncing, and also greater potential to catch larger prey, so why have the larger spiders abandoned them? The answer likely lies in the challenges of counteracting the force of gravity as their weight increases (see page 14). In order to bear the weight of a bigger spider, the silk threads of its web would have to increase in thickness more than proportionately. Thicker threads are costlier for the spider to produce, and are also easier for prey to see and therefore avoid. The largest spider that makes a web is *Nephila komaci*, a very rare species from Madagascar and South Africa that wasn't discovered until 2000. It's a lot smaller than the species we have discussed so far, with a leg span of 5 in (12 cm), but its web can reach an impressive 3 ft (1 m) in diameter.

Although *Nephila komaci* might be the largest spider that spins a web, it is not the spider than spins the largest web. Darwin's bark spider (*Caerostris darwini*) weighs about 0.5 g (less than the weight of a paper clip), and to build its web it first produces a silken line up to 80 ft (25 m) long stretching across a river. From this it hangs a web that is nearly 7 ft (2 m) in diameter, and that is renewed on a daily basis. This seems exceptionally expensive for the tiny spider, but the investment may reap rich rewards. The webs are only ever found across rivers or very close to them. Many insects—including dragonflies and mayflies—have a life cycle in which the immatures live underwater and the winged adults emerge *en masse* to mate and produce the next generation, and these mass emergences might lead to spectacular capture events for Darwin's bark spider. No one has observed this happening yet, but the species was only discovered in 2009, so we might just need a bit more patience.

Darwin's bark spider
Caerostris darwini
Web diameter: 7 ft (2 m)

This species builds the largest known webs, which are always over or near rivers to take advantage of flying insects drawn to water or emerging from it.

◀ Darwin's bark spider itself is not as impressive as its giant web would suggest, but its silk is the toughest biological material known, being ten times as strong as the Kevlar used in bulletproof vests.

Arthropod Adaptations

Insects and spiders are arthropods, characterized by having a tough external skeleton. We finish off our survey of this group by looking at the largest of all of the arthropods: crustaceans in the form of crabs and lobsters. However, to understand what both drives and limits large size in these animals, we first need to consider some general principles.

Arthropods on Land and Sea

The largest land-living arthropod comes in at around 9 lb (4 kg), while the largest recorded aquatic arthropod weighed 44 lb (20 kg), so it looks as though different effects limit the size of arthropods in the water and on land. This was demonstrated conclusively in 2015 in a large-scale study by an Anglo-Danish team of scientists. They studied twelve different groups of arthropod species from around the world in laboratory enclosures kept at different temperatures. They found that the body size of aquatic species generally reduced with warming water or decreasing latitude (moving away from the poles and toward the equator), but this effect was not seen in terrestrial species kept at warmer air temperatures. This suggests that oxygen is the limiting factor for aquatic species but not terrestrial ones.

The larger an animal is, the more oxygen it consumes. Furthermore, the higher the metabolism of the animal, the more oxygen it needs. Arthropods are ectotherms whose metabolic rate generally tracks the environmental temperature (see box), so their metabolism (and thus oxygen demand) increases when they are warmer. Temperatures are higher nearer the equator, and so oxygen demand here goes up. Warmer waters actually contain a little more dissolved oxygen, but this effect does not keep pace with the rate at which increasing temperature increases oxygen demand. So, it appears that the size aquatic arthropods can reach is limited by oxygen—indeed, the largest examples are found in very cold waters, where metabolisms are slowed and oxygen demand is reduced.

Endothermy and Ectothermy

Mammals and birds are endotherms, while almost all other animals are ectotherms. Endothermy is when body temperature is mostly maintained by the heat generated internally, whereas ectothermy is when the body is mostly warmed from external sources, mainly the sun. We humans eat a lot, and we burn up those calories in an active lifestyle, but our metabolic processes produce heat as a by-product and we use that heat to maintain a body temperature that is relatively constant and generally higher than that of our environment. In contrast, an ectotherm will seldom be warmer than its immediate environment—because it warms its body by gaining heat from the world around it—which is why we often see butterflies opening their wings and basking in the sunshine. By and large, the body temperature of a fish will be the same as the waters in which it swims, whereas a dolphin's body temperature will generally be higher than, and less affected by, the temperature of the surrounding water. This means that endotherms can remain active in cold conditions, but at the cost of having to find enough food to fuel their high metabolism.

Limits to Size

It might have occurred to you that fish are also ectotherms, so why can they grow to much larger sizes than aquatic arthropods? The answer is probably that fish (like all vertebrates) have a closed circulatory system that is much more efficient at delivering oxygen around the body than the simpler open system of arthropods. In a closed system, the blood is contained within a complex network of pipes and is pumped through this by the heart. This system allows fast and efficient movement of oxygen, and is responsive to the changing needs of different body parts—for example, we send much more blood to our muscles when we are exercising. In contrast, in an open system the heart

▲ The Japanese spider crab (*Macrocheira kaempferi*) has the longest leg span of any arthropod—16 ft (5 m).

◀ Weighing in at 9 lb (4 kg), the coconut crab (*Birgus latro*) is the world's largest terrestrial arthropod. Its maximum size is likely limited by the weight of its exoskeleton.

▶ At 44 lb (20 kg), the American lobster (*Homarus americanus*) is the heaviest arthropod alive. It should be handled with care, as it will not hesitate to use its fearsome claws when it feels threatened.

◀ Giant spider crabs (*Leptomithrax gaimardii*) are known to pile up together in huge groups at spots around Australia when they molt. It may be that they gain safety in numbers during this time, when they are most vulnerable to predation.

simply pumps blood into all the body cavities, from where it percolates through the tissues and eventually finds its way back to the heart again. Open systems work well in small animals where the blood never needs to travel long distances, but the larger the animal, the more valuable the extra control of a closed system. Thus, aquatic arthropods are limited in size by the ability of their open circulatory system to deliver enough oxygen to fuel their metabolism.

Weighty Issues

Crabs are just one group of arthropods that can become larger in the water than on the land, so there is clearly some restriction to life on land for arthropods that kicks in before the issue of oxygen delivery. This restriction is likely linked to their exoskeleton. Compared to our thin skin, an exoskeleton provides superb protection from injury and from attack by parasitoids and predators. But this comes at a cost. The exoskeleton is like a stiff suit of armor, and as the organism grows it ends up being too tight, such that the animal has to burst out and then grow a bigger one.

Each molting episode costs the organism in a number of ways. The process of shedding the exoskeleton is energetically demanding and sometimes goes wrong—a common source of mortality among arthropods is getting trapped in their suit of armor with only half of it shed. The period when the organism regrows a new exoskeleton and lets it harden is also energetically expensive, and it leaves the animal unable to go about its normal activities—like looking for food—and makes it very vulnerable to predators. All these costs increase with size: a larger suit of armor requires more resources to produce, takes longer to make, takes longer to harden, and is more difficult and time consuming to peel off. However, all these costs probably apply just as equally in water as in air, so none is likely to be the factor that limits size on land more than in water.

Another drawback to having a suit of armor is the energetic cost of moving around in it—simply put, an exoskeleton is heavy, and the animal requires considerable leg power to stand up and walk around in it. As we saw in Chapter 1, much of the weight of an organism in water can be counteracted by buoyancy at no energetic cost, and the excess muscle power available to an organism to carry heavy loads declines with its size. Hence, it seems likely that it is an arthropod's declining ability to bear the weight of its exoskeleton that finally limits its size on land. Some of the largest land arthropods are millipedes (see box opposite and page 152), which have plenty of legs to bear their weight.

▲ A lobster emerging from its shed carapace during molting—its exterior will soon darken as it hardens.

Millipede or Centipede?

Counting the total number of legs of a millipede or centipede isn't a reliable way to tell the two apart. For a start, although the word millipede is Latin for "1,000 feet," no millipede has that many limbs—the largest recorded number was 750—and an individual can change the number of legs it has when it molts. The most reliable distinction between millipedes and centipedes is that while both have long bodies made up of a line of similar segments, a millipede has two sets of legs emerging from each segment and a centipede only has one set. If checking for this character involves getting closer to the beast than you would like, then remember that centipedes generally run faster and prey on small animals rather than eating plant matter.

▼ The impressive size and docile vegetarian nature of giant millipedes have made them popular pets.

▼ This large centipede likely has bright coloration to warn would-be avian predators that it is toxic to eat.

King-sized Crustaceans

Let's switch from looking at general principles limiting arthropod size to describing some enormous examples of these animals. To do this, we need focus only on the crustaceans—in particular the crabs and lobsters—where we find both terrestrial and aquatic giants.

Coconut crab
Birgus latro
Weight: 9 lb (4 kg)
Coconuts provide the main part of this species' diet, together with a protective home that it can carry about.

The Coconut Crab

The coconut crab (*Birgus latro*) is the largest terrestrial arthropod, with a leg span up to 3 ft (1 m) and a weight of around 9 lb (4 kg). It is found across the Pacific and Indian oceans, where its distribution closely matches that of the coconut palm (*Cocos nucifera*). The adults are obligate land dwellers that easily drown in water, but the females lay their eggs in the sea and the young spend their first month feeding in the water column. It is assumed that the crabs colonize new places by hitching a ride on floating objects, giving them a means of escape from fish predators and from the energetic costs of swimming upward against the force of gravity to avoid sinking to the bottom. Coconuts readily float and are known to

travel long distances intact, so the close association between crab and tree might be because the two often disperse together. But this is not the only way the crab uses coconuts.

When coconut crabs are small, they don't invest heavily in an exoskeleton and instead take the hermit crab option for protection, living in a discarded snail shell. As they get bigger, sufficiently large shells become harder to find. Eventually, the biggest individuals give up the hermit lifestyle and form a tough outer skeleton, while intermediate-sized individuals often use coconut-shell fragments as an alternative—if you see a coconut moving mysteriously across a Pacific beach, this is likely the explanation! Adult coconut crabs are omnivores, but coconuts make up a large part of their diet. Depending on the size and ripeness of the coconut, they can cut a hole in its shell with their strong claws and scoop out the flesh inside (and in so doing make ideal temporary homes for smaller crabs once they have finished their meal). However, some coconuts are too tough for the crabs to break into with their claws, and they carry these with them as they climb a tree, then drop the fruit from a height in the hope that it breaks open on a rock below. The crab can then climb down and enjoy the contents—although they often save the time and effort of climbing down by just letting themselves go and falling. A fall that can split a coconut open leaves the crab undamaged, which is proof of the fabulous protection an exoskeleton offers. All in all, the coconut crab is pretty exceptional, and few species are more aptly named.

The Japanese Spider Crab

The Japanese spider crab (*Macrocheira kaempferi*) is also aptly named, as it has exceptionally long, spider-like legs with a span of more than 16 ft (5 m), making it the biggest (if not the heaviest) crab. However, the species has been subject to intensive fishing, and individuals don't generally survive fishermen's nets

Japanese spider crab
Macrocheira kaempferi
Leg span: 16 ft (5 m)

Despite its large size, the Japanese spider crab is well camouflaged, making it quite difficult to spot against the background of a coral reef. The crustacean sometimes enhances this disguise by fastening pieces of sponge to the outside of its body.

long enough to grow that big—most of those caught span only 3 ft (1 m) or so. You wouldn't think that such a heavily armored giant had much to fear from any predators other than humans, but apparently this is not to the case—why else would these crabs be such masters of camouflage? Not only does their lumpy, bumpy shell offer a good match for the rocky ocean floors on which they sit, but the crustaceans actively enhance their disguise by sticking bits of coral and sponges to their back. The crab's behavior is unlikely to be concerned with camouflaging itself to sneak up on prey, as most of its diet comprises vegetation or animals like shellfish that can't run way. Instead, it is aimed at hiding itself from predators, the most likely of these being octopuses, which are generally very strong and can sometimes be pretty giant themselves (see page 160).

The American Lobster

The last of our giant crustaceans, the American lobster (*Homarus americanus*), should also be the most familiar. It is harvested along the entire northeastern coast of North America, so if you treat yourself to lobster in a fancy restaurant in the US or Canada, this is the species that will be delivered to you. Although the American lobster can't touch the Japanese spider crab in terms of size, it is the heaviest of the crustaceans, reaching weights up to 44 lb (20 kg). If your fancy restaurant is in Europe, then you will most likely be served a close relative of the American lobster—the

European lobster (*Homarus gammarus*). The two species are so similar, in fact, that they can interbreed if placed in a tank together. Since the American lobster is so familiar, we won't dwell on it, suffice it to say that its sex life also highlights another issue with wearing a suit of armor: mating can occur only when the female periodically molts her hard exoskeleton.

American lobster
Homarus americanus
Weight: up to 44 lb (20 kg)

Cooking causes the carapace of the American lobster to turn a more uniform vivid red color.

Ancient Arthropods

Consideration of giant prehistoric arthropods gives us a chance to emphasize that our knowledge of extinct animals can sometimes be based on little physical evidence and a lot of logical reasoning and educated guesswork. Here, we consider a clawed aquatic predator before turning to giant millipedes. Although the sizes quoted for these extinct giants are nothing like those of the Japanese spider crab (see page 150), most of that species' span is down to its long, spindly legs, with its body making up perhaps only 15 in (40 cm) of the 16 ft (5 m) total. In contrast, the extinct giant arthropods we look at here had much bigger bodies and so were much heavier.

Pincered Predator

Swimming predatory arthropods of the genus *Jaekelopterus* are thought to have lived about 400 million years ago. Members of the genus are estimated to be the largest arthropods ever known, with a length of about 8 ft (2.5 m), based on a partial giant claw measuring 13 in (34 cm) in length. If we assume that these animals had claws of a similar shape to smaller close relatives, then we can estimate that the full length of the claw was 18 in (46 cm). And if we further assume that the ratio of claw length to whole body length was similar in this genus to that of closely related smaller relatives for which we have more fossil material, then we can estimate its length as being 8 ft (2.5 m).

The above assumptions aren't too fanciful, because when we look at a range of closely related species for which we do have good fossil evidence, then we find that they are pretty consistent in body shape and claw shape. Nevertheless, they are assumptions—the problem for us is that although the claws of these giants were very robust and have fossilized well, it appears the animals had a very lightly armored exoskeleton that was readily destroyed after death and so is unlikely to be found as a fossil. Given that the animals were probably swimmers (like their smaller modern-day relatives), the lighter they could be, the less energy they would expend in swimming. And given that they were large and had terrifying claws, they could probably protect themselves very well from most predators by fighting them off, and therefore would have had no need for a tough (and heavy) suit of armor. However, many types of animals would have been able to break through the light exoskeleton following the death of such an arthropod and scavenge its corpse, in the process ripping its

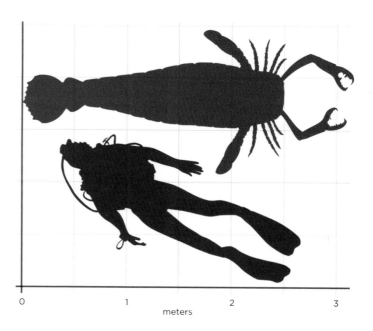

▲ Sea scorpions (genus *Jaekelopterus*) were the largest known arthropods, measuring up to 8 ft (2.5 m) in length— considerably longer, and heavier, than an adult human.

body to shreds and leaving only the indigestible heavy claws to sink down into the sediment and fossilize over time. Even if the estimated length is a little on the high side, these might well still be the biggest-bodied arthropods ever, since they were a good 20 in (50 cm) longer than their nearest rivals—a genus of ancient millipedes.

Many-legged Monster

Millipedes have been around for the last 430 million years and there are 12,000 named living species, most of which are slow-moving consumers of dead plant matter. They can't bite you, but they can inflict a painful sting by secreting noxious chemicals along the body as a main form of defense.

▲ The primary function of a sea scorpion's giant claws was likely to catch fast-moving prey. The marine arthropods may also have been strong enough to crush some shelled prey, and would have been able to defend themselves against many predators.

The largest living millipede is the giant African millipede (*Archispirostreptus gigas*), which is widespread across low-lying parts of East Africa. It can grow to 15 in (39 cm) long and 2⅗ in (6.7 cm) wide, and has about 256 legs. The millipedes are quite popular as pets, especially as they are willing to eat pretty much any fruit and vegetables in any condition—they are a novel way of recycling a lot of kitchen waste. However, these arthropods are positively dwarfed by a genus of giant millipedes that lived about 300 million years ago in northwest North America and Scotland

(which were joined at the time). The largest of these grew to 7.5 ft (2.3 m) long and 20 in (50 cm) wide thanks to the high oxygen levels in the atmosphere at that time and a relative lack of vertebrate predators large enough to attack them. But they did not last, as the lush forests of the Carboniferous period retreated in the much drier Permian, and their food supply dry up. There are many reasons to visit the beautiful Isle of Arran of the west coast of Scotland; if you do so, be sure to head to Laggan Harbour to see a perfectly preserved set of footprints from this giant.

▲ The African giant black millipede (*Archispirostreptus gigas*) can have as many as 256 legs, although this number varies between individuals and even over an individual's lifetime.

▲ Giant extinct millipede tracks from the Late Carboniferous period, now exposed on rock on the Isle of Arran off the west coast of Scotland.

Gelatinous Giants

One cost to becoming a giant is that you have to eat more in order to meet the increasing metabolic demands of your body. One way around this is to grow big by carrying a lot of something that isn't costly to maintain—like water. Jellyfish are 95–98 percent water, and some of them are huge. This giant size gives them the benefit of being able to sweep large volumes of water in search of prey, but without having to pay high metabolic costs.

▲ Jellyfish come in a stunning variety of sizes, shapes, and colors.

Jellyfish Overview

Jellyfish aren't fish at all, as fish have a backbone and, being invertebrates, jellyfish do not. Hence, some scientists prefer to call them jellies rather than jellyfish, or, more formally, members of the subphylum Medusozoa of the phylum Cnidaria. Since we are not overly concerned with careful taxonomy here, we will stick to calling them jellyfish. Regardless of what you call them, you would recognize one if you saw it washed up on a beach: as adults, they are soft-bodied free-swimming aquatic animals characterized by a gelatinous umbrella-shaped bell and often a large number of trailing tentacles. They move by pulsing their bell, and they use their tentacles in defense and to capture prey. Your parents likely taught you never to touch any jellyfish you see on the beach because they can sting even after death.

Jellyfish Ecology

We seem to be experiencing a period of growth in jellyfish numbers worldwide. This can impact humans adversely in a number of ways: through the effect on tourism if high numbers of jellyfish on coasts make recreational activities in the water painful or even dangerous; by clogging fishing nets, power-station outflows, and ship propulsion systems; and by killing farmed fish.

The reasons for the population increases are unclear. One suggestion is that overfishing is an important cause, the reasoning being that the fish species humans like to eat often compete with jellyfish for the same food. A wrinkle in this story is the suggestion that if we stop targeting a particular fish to allow it to recover, the growth in jellyfish numbers we previously facilitated could hinder fish stock recovery because they eat the very early stages of the fish.

In general, jellyfish can thrive in more nutrient-poor waters than fish, which leads to two other possibilities for why jellyfish numbers are growing: climate change and the increased use of nitrogen fertilizers. As global temperatures rise, so do sea surface temperatures. This tends to stop mixing of the ocean layers, so that as the abundant life in the upper layers uses up the available nutrients, these are not replaced from deeper waters. Similarly, as farmers add large quantities of nitrogen fertilizer to land, some of this is washed into rivers and out to estuaries, where it can trigger a plankton bloom that soon exhausts all the available nutrients. These are not the only possible reasons for increases in jellyfish numbers, and the relative importance of different factors will vary. But no matter where you are in the world, if you are walking along the beach you should expect to find more stranded jellyfish than was once the case, and among them you might even come across some giants.

A Rare Giant

One jellyfish species you would be very lucky to see is *Stygiomedusa gigantea*. The bell of this species is more than 3 ft (1 m) wide and its tentacles are 33 ft (10 m) long. It has been recorded only about a hundred times in the last century, so you are unlikely to come across one—although sightings have been right across the globe, so you are in with a chance wherever you happen to go swimming. One mystery is how this giant feeds, as it has an unusually small number of tentacles (four) and no stings. My guess is that it uses its four arms to grab hold of unusually large prey, restraining the animal with the strength of its grip, and then lifts that prey toward its digestive organ in the bell. However, this remains no more than a wild guess on my part, and we will just have to wait until an actively hunting individual wanders into the viewer of someone's underwater camera.

▲ The low metabolism of jellyfish allows them to exist on a meager food supply at very high densities.

◀ A jellyfish's low metabolism can be appreciated when you consider just how much of its body is water; this can also sometimes make the animals difficult to see.

Stygiomedusa gigantea
Length: 33 ft (10 m)

This amazing giant jellyfish was photographed by the Monterey Bay Aquarium Research Institute in the Gulf of California.

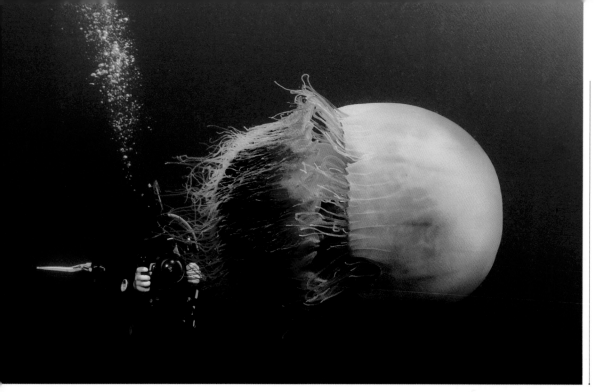

Nomura's jellyfish
Nemopilema nomurai
Weight: 440 lb (200 kg)

Its huge size makes this species very slow, allowing divers to keep pace with it easily.

An Increasingly Common Giant

Nomura's jellyfish (*Nemopilema nomurai*) has a bell measuring up to 6.5 ft (2 m) across and can weigh 440 lb (200 kg). It is found in the waters between China and Japan, where blooms of giants are becoming increasingly common. When swarms occur, fishing becomes impossible as nets fill with the jellyfish, and there is some concern that the jellyfish will target larger fish as they get bigger themselves, putting them in direct competition with human fishermen. We are used to giants that live for decades, but jellyfish have a very short life cycle—few live longer than six months. However, Nomura's jellyfish may lay claim to the fastest-growing animal, reaching three times my weight just six months after it emerges from the egg as something smaller than a grain of rice!

The Largest of Them All

The lion's mane jellyfish (*Cyanea capillata*) has a very different body plan from Nomura's jellyfish, with a tangle of countless fine stinging tentacles that is reminiscent of the mane of a male lion—hence its common name. It is found in cold waters around the Arctic and North Atlantic, and perhaps also around Australia and New Zealand (although the huge separation may indicate that the southern population is actually a separate species). The bell on the largest individual found was 7.5 ft (2.3 m) across and some of the tentacles were 121 ft (37 m) long. To put that in perspective, a large blue whale (*Balaenoptera musculus*; see page 76) might be 82 ft (25 m) long and a Boeing 737 is about 115 ft (35 m) long. This jellyfish may therefore possibly be the longest animal alive (although there is a worm that might just beat it—see page 168).

The lion's mane jellyfish is so big that it provides a unique habitat for small fish and other marine animals—if they are immune to the stings or are nimble enough to dodge them, then the mane is an attractive place in the featureless ocean to hide from large predatory fish. Some fish may even be able to get all the food they need from the crumbs dropped by the jellyfish. Regardless, these jellyfish often have a menagerie of dozens of different types of animals taking cover under their mane.

◄ Nomura's jellyfish periodically occur in dense swarms, choking the nets of local fishermen.

Fossil Jellyfish

Jellyfish have been around for at least 500 million years, so it is very possible that there were larger species in the past. However, you won't be surprised to hear that we don't have extensive fossil records of jellyfish—conditions have to be very unusual for such a fragile, soft-bodied animal to fossilize well. So far, scientists haven't found evidence of species that were bigger than the largest we see today, and my guess is that we won't find any that are substantially larger than the lion's mane jellyfish. In general, jellyfish are energy-efficient yet sluggish swimmers, which is why they can get caught out by changing currents and tides, and end up stranded on beaches—consider how often you find jellyfish washed up compared to fish. The larger a jellyfish gets, the slower it is able to contract its bell and so the more sluggish it gets. The lion's mane jellyfish is a slow swimmer and not terribly maneuverable, and a species that is even bigger would be too slow and clumsy to be able to locate and track good feeding areas and to avoid predators. Even the giant lion's mane jellyfish has predators when fully grown—if it nears the surface, seabirds will attack it, and it also seems to be a favored food of the giant ocean sunfish (*Mola mola*; see page 86) and the largest of the turtles, the leatherback (*Dermochelys coriacea*; see page 184).

Flying Jellies?

We don't see non-aquatic animals getting to large sizes on the cheap by filling out with a metabolically low-cost material as do jellyfish. We wouldn't expect a water-filled terrestrial organism to evolve because it would simply be too heavy, but how about a gas-filled one? It would be unlikely for a terrestrial animal to have a large gas balloon because the thin membrane would be too vulnerable to puncture from sharp stones, thorns, and the like, but could we imagine a living airship? There is not, nor has there ever been, such a thing. The likely reason for this is partly also why airships have been superseded by airplanes. Airships are fuel efficient, but they are slow and they can very easily get blown off course by the wind, thus a biological airship would struggle to find and track the food it needed. That food would be an issue too, as the density of food in the air is nothing like that of the food available in the surface waters of the oceans or on land. Birds that make a living catching insects on the wing have to be nimble enough of seek out short-term aggregations of prey, which is not something a sluggish biological airship could manage.

▲ The lion's mane jellyfish (*Cyanea capillata*) can flourish in cold waters—this individual was photographed off Scotland.

▼ The lion's mane jellyfish might be the longest animal alive, with the tentacles of some individuals reaching 121 ft (31 m) in length.

Colossal Squid

It will come as no surprise that the largest living invertebrate—the colossal squid (*Mesonychoteuthis hamiltoni*)—is aquatic. The largest individual that has been measured came in at nearly 1,100 lb (500 kg), but there is good reason to think that there are even bigger ones swimming the waters of the Arctic Ocean. These creatures seem to be an important component of the diet of sperm whales (*Physeter macrocephalus;* see page 91), but there is evidence that they can put up quite a fight against the biggest predatory whale.

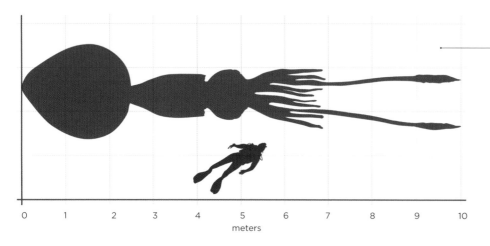

0 1 2 3 4 5 6 7 8 9 10
meters

Colossal squid
Mesonychoteuthis hamiltoni
Weight: up to 1,091 lb (495 kg)

Although this species is the largest living invertebrate, it weighs less than a cow—indicating that animals require a skeleton to reach a giant size.

Full Measure

The colossal squid was first described in 1925 on the basis of two tentacles inside the stomach of a sperm whale from a species never before seen. It was not until 1981 that a whole animal was captured in the net of a fishing trawler, and to date we have only ever found about a dozen complete individuals. The largest one caught so far by fishermen weighed 1,091 lb (495 kg) and is now on display in Te Papa, New Zealand's national museum in Wellington. However, we have reason to believe that there are even bigger individuals waiting to be found.

The least digestible part of a squid is its beak, a hard mouthpart like a bird's bill that is used for slicing up prey. We know which marine predators eat different species of squid by examining the beaks found intact in their stomachs. Colossal squid beaks measuring 2 in (49 mm) long have been found in sperm whale stomachs, but none of the beaks in individuals that have been captured whole has been that size—the longest was 1⅗ in (42.5 mm), from the 1,091 lb (495 kg) individual. Hence, it is likely that there are even bigger colossal squid than the ones we have found, but how big an individual with a 2 in

▲ The tough beak of a squid, used to cut up food, is tiny compared to its body—this one belongs to a colossal squid.

▲ This is the largest colossal squid ever found, now preserved in Te Papa, New Zealand's national museum.

(49 mm) beak might be is hard to guess because the relationship between total size and beak size is not simple. It's a bit like if we found the lower jawbone of a person, in that this would allow us to estimate the size of the person to some extent (for example, whether they were an adult or a child), but not their exact height or weight. My weight has decreased by nearly 20 percent in the last few years, but my jawbone has not changed in size at all.

We can, however, be sure that the colossal squid really is the biggest invertebrate. The second largest is also a squid—the giant squid (*Architeuthis dux*). Nearly a thousand of these creatures have been caught, so scientists are pretty confident that we aren't going to find one much bigger than the existing record-holder at 650 lb (295 kg)—and that's a lot less than the colossal squid found to date.

The Hunter and the Hunted

What hunts colossal squid and what do they eat themselves? The first part of this question is easy to answer thanks to the beaks of these animals. Beaks of immature colossal squid have been found in the stomachs of a whole range of whales, southern elephant seals (*Mirounga leonina*), some large fish, and even albatrosses, but those of really large individuals have been found only in sleeper sharks and sperm whales. Sleeper sharks can reach lengths of 15 ft (4.5 m) and weigh more than 1,760 lb (800 kg), but even so, a healthy, full-grown colossal squid wouldn't be an easy meal. The recorded instances of large beaks found in sleeper sharks might be explained by the sharks only occasionally targeting sick or injured individuals. However, large beaks are regularly found in sperm whale stomachs, and it seems that they form an important part of the whale's diet. There is a theory that sperm whales can stun their prey with loud echolocation clicks, but this is definitely not true for squid. Sperm whales often have scars on the skin around their head, which are often such perfect impressions of the suckers on squid tentacles that the species of squid that left the marks can be identified. Colossal squid feature heavily in these battle scars.

We know that sperm whales start these fights. Colossal squid likely mostly eat members of their own kind (cannibalism is common throughout different squid species) as well as the two toothfish species that are relatively common in the cold Southern Ocean. It has been suggested that colossal squid are normally sluggish ambush predators that expend little effort in searching for food, and this—combined with a low resting metabolism in the cold waters of the Southern Ocean—indicates that their food requirement might be as low as 66 lb (30 kg) of fish a day for even a 1,100 lb (500 kg) individual. That is equivalent to me making do on just a single can of tuna every three weeks!

Octopuses

Since earliest times there have been tall tales of octopuses pulling people off boats or even giants dragging whole boats down. Sadly, however, these stories aren't backed up by any real evidence. The only octopuses of the 300 known species that might potentially kill a person are the four blue-ringed species of the Pacific and Indian oceans, but these are not giants (they have a span of just 6 in/15 cm) and instead are deadly thanks to their venom.

▲ The heaviest recorded giant Pacific octopus (*Enteroctopus dofleini*) weighed 157 lb (71 kg).

Octopuses or Octopi?

The main controversy surrounding the octopus is whether the plural is octopuses or octopi. I have used octopuses here, but you will often see octopi in books and online. You will hear some people say that octopi is wrong because the word octopus derives from Greek rather than Latin. I am a biologist, not a linguist, but it seems to me that English is full of exceptions and anomalies, and nothing that is as commonly used as octopi can really be labeled "wrong." You could adopt the rare plural octopodes as well, but only if you are the sort of person who asks if they can borrow someone's "cell telephone."

Sea Monsters

The largest octopus is the giant Pacific octopus (*Enteroctopus dofleini*). Fishermen harvest about 3 million tons of these each year for human consumption, so you would think we would know exactly how big they get. The fishermen would say that 33 lb (15 kg) is a good-sized adult, but certainly a number of individuals weighing more than 110 lb (50 kg) have been documented. The biggest one for which we have good evidence weighed 157 lb (71 kg), but if you browse the non-scientific literature you will find values of 300 lb (135 kg), 400 lb (180 kg), and 600 lb (270 kg) quoted. I am not saying that these sizes are impossible, just that there isn't a preserved specimen to back them up, or even a reliable thread of documentary evidence leading back to an unbiased observer who actually weighed the individual with an appropriate set of scales.

The only other species that comes close to the giant Pacific octopus in terms of size is the seven-arm octopus (*Haliphron atlanticus*). Some fishermen did find an incomplete individual of this species in their net that weighed 134 lb (61 kg), and we know enough about the morphology of this species to make a pretty good guess of the weight of the entire animal when it was alive, and that estimate comes out at 165 lb (75 kg). So, you could argue endlessly about which of the two species is actually the bigger, or you could just accept that the very biggest individuals of both species are about as heavy as an adult man and get on with the rest of your life.

Giant Pacific octopuses are popular in public aquariums on account of their impressive size and generally active nature. In my younger years I regularly visited an aquarium that provided a child's construction set for its octopus to play with, and while the animal never built a scale model of the Eiffel Tower, it was always turning over the pieces and

▲ Pressure from human hunting means that most giant Pacific octopuses don't get the chance to grow much bigger than this (still impressive) individual.

examining them. Octopuses frequently escape from their tanks for a number of reasons. For a start, many species can survive out of water for several minutes. Second, they are really strong—a single large sucker on one of the eight arms of a large giant Pacific octopus can lift a weight of 35 lb (16 kg). Third, they are really flexible: their only incompressible body part is their beak, so they can fit their whole body through a hole the size of their beak—and the beak of a 110 lb (50 kg) individual might be only 2–2⅖ in (5–6 cm) across. Finally, they are considered to be very intelligent. Scientists have to take great care to minimize suffering when carrying out experiments on any animal, but generally they are required by legislation to take exceptional care with vertebrates since these usually have more complex nervous systems. In the UK and Canada, octopuses are the only invertebrate species that are listed as requiring the same special treatment as the vertebrates.

▲ Based on partial remains, the seven-arm octopus (*Haliphron atlanticus*) is estimated to reach a weight of 165 lb (75 kg). This unfortunate individual is being attacked by a loggerhead turtle (*Caretta caretta*).

◀ While the four species of blue-ringed octopuses are not giants, they are the most highly venomous marine animals.

More Massive Mollusks

In this section I would like to introduce and scotch some myths about another massive aquatic invertebrate, the giant clam (*Tridacna gigas*). The clams are distant relatives of the squids and octopuses, being grouped together with those animals in the phylum Mollusca—invertebrates that have (or had at one point in their evolutionary past) a protective shell. We also have a look here at some extinct members of the mollusk lineage in order to highlight general trends in giant prehistoric invertebrates.

The Giant Clam

Found on shallow-water reefs in the Indian and South Pacific oceans, the giant clam can exceed 3 ft (1 m) in length and the largest individuals ever found weighed 550–660 lb (250–300 kg). Although stories abound of these creatures trapping humans, we will examine here why these are pretty implausible, as well as why the mollusks have much more to fear from us than we do from them and how a single individual can produce something worth more than $100 million.

Until quite recently, the largest known giant clam was a specimen found off the coast of Sumatra, Indonesia, in 1817, which was estimated to weigh 550 lb (250 kg). In my lifetime there have been another couple of finds that might be a little bigger, although there has been no definitive study of these. If I were to ask you to guess where you might go to see the preserved shells of the 1817 Sumatran giant, it would probably take a lot of attempts before you arrived at the Ulster Museum in Belfast, Northern Ireland. It seems that the shells were donated to the Belfast Natural History Society by the English entomologist Francis Walker in the nineteenth century. How he obtained them isn't clear, although he worked for thirty years for the Natural History Museum in London, which would have brought him into contact with a lot of commercial collectors selling unusual natural history specimens. Walker doesn't seem to have had a particularly strong connection with the Belfast society but, like the naturalist Charles Darwin, he was an active corresponding member, and he dispersed his private collections to several museums and learned societies toward the end of his life.

Clam shells fossilize well, and from finds of these scientists know that there were even bigger species in previous eras. *Platyceramus platinus*, which lived during the Cretaceous period, resembled today's giant clams but could have shells that were three times as long and perhaps six to ten times as heavy.

▲ Extensive human exploitation is endangering the giant clam (*Tridacna gigas*). Even if the species does not go extinct, we may never see individuals that survive long enough to equal the largest recorded individual, which weighed 660 lb (300 kg).

▲ The strong zigzag pattern of the giant clam's shell can be seen clearly in this example.

The Human Threat

Giant clams are in fast decline, to the extent that the species is now extinct in many of its former strongholds. The problem here is overharvesting by humans—the flesh of the clams is considered a delicacy in some parts, their shells can be turned into stunning ornaments, and the adductor muscle the animals use to close the two halves of the shell is considered by some to have aphrodisiac powers. The disparity between what rich foreign buyers will pay for them and the very low incomes of many locals makes conservation a huge challenge—not unlike that faced by the African bush elephant (*Loxodonta africana*; see page 42).

There have been accounts of divers becoming trapped and drowning when the two shells of a giant clam snapped shut and held them fast by an arm or leg, but these sound like tall tales to me. During daylight, a giant clam will hold the two parts of its shell wide open so that it can feed. In fact, most of the animal's nutrition comes from the symbiotic relationship it has with photosynthesizing algae living in its flesh. The algae leach some of their nutrients into the flesh of the clam, which absorbs these and in return provides the algae with a safe home. This is why giant clams live in shallow water where there is plenty of sunlight, and also why they can grow to a large size even in nutrient-poor waters. Of course, their dependency on shallow waters makes them easily accessible to human divers. If a diver did start poking around in a giant clam, then the clam would respond by closing its shell. However, as the larger individuals take 30 seconds to close and in any case are unable to close fully, it seems unlikely that they would trap a limb.

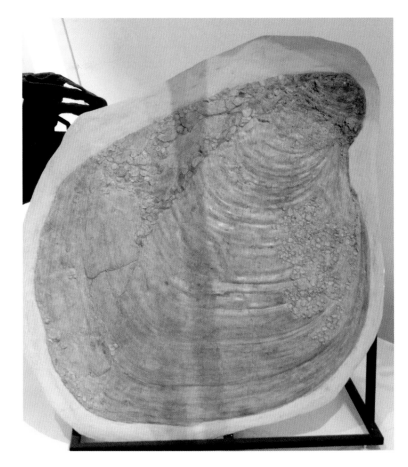

▲ *Platyceramus platinus*, from the Cretaceous period, is the biggest bivalve mollusk discovered to date.

▲ The shell of the biggest giant clam ever discovered, now on display in the Ulster Museum in Belfast, Northern Ireland.

Giant Pearls

All bivalve mollusks (those that protect themselves with a pair of shells), including giant clams, produce their own shells, growing them as they themselves grow. If a foreign body such as a grain of sand gets inside the shell and becomes lodged in a position where it can't be washed out, then the mollusk secretes material around it like that used for the inside of its shell in a bid to make it less irritating as it rubs against its flesh. The animal may end up secreting many layers of this material, creating the pearls we covet for jewelry. Some bivalves do this more readily than others, and incidences are very uncommon in giant clams, but in 2006 a Filipino fisherman found a 75 lb (34 kg) giant clam pearl—the biggest pearl ever. He kept it under his bed as a good luck charm without telling anyone, until eventually he handed it over to the local mayor in 2016. Like works of art, pearls are worth only what someone is prepared to pay for them, but based on the sums paid for other large pearls, this one might be worth $100 million. The global publicity surrounding this find will not be a great help to giant clam conservation, so I should reiterate here that giant clam pearls of any size are very rare indeed.

▲ A bivalve mollusk opened to reveal a pearl.

▶ A huge (and thus extremely valuable) pearl, found in 2006 inside a giant clam in the Philippines.

Spiraling Up

It might seem strange that octopuses and squid are grouped along with all the shelled organisms as mollusks, but they derived from an ancestor that had a prominent shell. Very closely related to the squid and octopuses are the six living species of nautilus. These animals swim through the water looking for food in the manner of squid, but they have a spiral shell. The shell is made up of a series of compartments of increasing size, and every now and again the nautilus builds a bigger compartment at the end of its spiral, moves into these more spacious quarters, and seals off its previous living area. The previously occupied chambers are partly filled with gas, which explains why an animal with such a heavy shell can still swim. When the animal is feeding, a considerable portion of its body extends from the

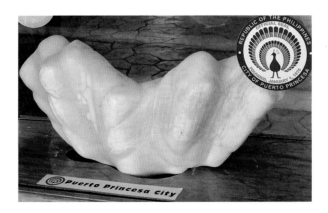

opening of its living chamber, but if it is threatened, it can retreat entirely inside and block the entrance with two specially toughened coiled tentacles. This offers nautiluses great protection, allowing them to live for more than twenty years when squid and octopuses of the same size rarely live longer than two years.

Nautiluses are not huge—the largest individuals have a maximum shell diameter of 10 in (25 cm)—but they have remained virtually unchanged for 500 million years. For much of that time they shared the sea with another group of shelled mollusks, many of which produced similar spiral shells: the ammonites. Some of these were really huge—for example, incomplete fossil remains of the spiral shell of one *Parapuzosia seppenradensis* individual measured 6 ft (1.8 m) across. The final (largest) chamber of this animal's spiral is incomplete, and it is estimated that the original shell might have been 8–11.5 ft (2.5–3.5 m) across and weighed about 1,550 lb (700 kg), with the occupant weighing another 1,650 lb (750 kg). Huge numbers of ammonite fossils have been found, but these marine mollusks became extinct during the Cretaceous-Paleogene extinction event 66 million years ago that also finished off the dinosaurs. One possible reason why nautiluses survived but ammonites did not is that nautilus juveniles spend their time on or near the ocean floor, whereas juvenile ammonites are thought to have fed in surface waters. When the meteor struck and wiped out much life on Earth, the depths of the ocean were probably buffered from the adverse effects to a greater extent than the surface waters.

▲ A nautilus is mostly concealed by the outer chamber of its shell.

◀ A cross section of a nautilus shell, showing its spiral shape and multiple chambers.

▶ An ammonite in the genus *Parapuzosia*, which included the largest individuals in the group—some had shells reaching 11.5 ft (3.5 m) across.

Weighty Worms

Worms have such a simple body plan that it should be easy for them to grow to spectacular lengths. For an earthworm, being larger might allow it to hold earth in its digestive system longer and so extract nutrients from even the poorest soils. Also, being bigger may give it greater strength to burrow through very compacted soil. At the same time, we would expect the size of earthworms to be limited, as the bigger an individual is, the more friction it will experience against the soil as it drags itself along. A number of impressively long earthworms do exist: the Australian giant Gippsland earthworm (*Megascolides australis*) can reach up to 10 ft (3 m) in length, while the African giant earthworm (*Microchaetus rappi*) can reach 23 ft (7 m). However, the biggest worms are not actually earthworms.

A Whale of a Worm

If you want to find a really big worm, then you have to look inside the gut of a large mammal. This is a likely location because mammals are endothermic, so their gut and any parasite within it are kept at a temperature conducive to digestion, and the high eating rate of mammals means that parasites living in their gut are often awash with potential food, which they absorb through their skin. Best of all, the parasite doesn't need to move to find food, but simply has to anchor itself within the intestines (to avoid being pushed through and excreted) and let the host do all the hard work. If a parasite can remain undetected by the host's immune system, then it has plenty of room to grow—the small intestine of an adult human, for example, is about 23 ft (7 m) long. If you were looking for the biggest parasitic worm, then a pretty likely place would therefore be inside the biggest mammals—and indeed the tapeworm *Polygonoporus giganticus* is found in the second-largest whale species, the sperm whale (see page 88), and reaches lengths in excess of 100 ft (30 m).

Despite weighing up to 63 tons (57 tonnes), sperm whales are only a third the size of the largest living animal, the blue whale (see page 76), so it wouldn't seem unreasonable to expect to find even bigger intestinal parasites in this species. However, this doesn't appear to be the case. Humans have caught and butchered literally hundreds of thousands of blue whales, and so if the animals harbored giant gut worms then these would have been found and reported. What saves blue whales from huge parasites is probably their diet. When tapeworm eggs are shed into the environment, they are taken up by a smaller

intermediate host, where they remain in the tissues of the animal until it is eaten by the ultimate host. Here, the parasite develops into the mature form, which produces the next generation of eggs. In the case of *Polygonoporus giganticus*, it is likely that a large fish serves as the intermediate host. These animals might live for more than a decade, giving them plenty of time to pick up parasites as they feed, and they also have a fair chance of being eaten by something like a sperm whale, in which the parasite can develop. In contrast, blue whales eat tiny crustaceans called krill, which are only 2/5 in (1 cm) long. These animals don't live very long and most of them probably don't end up being eaten by a blue whale, so it is difficult for a parasite to rely on blue whales for part of their life cycle.

The average adult male sperm whale weighs about 44 tons (40 tonnes), or about 500 times more than the average adult male human at 175 lb (80 kg). If a sperm whale tapeworm can grow to 100 ft (30 m), then what would be your best guess for the length a human tapeworm might reach? Surprisingly, it isn't that much smaller. If you eat undercooked beef, then there is a risk you may be infected with the beef tapeworm (*Taenia saginata*); this commonly grows to 13–33 ft (4–10 m), but the record stands at 72 ft (22 m). So, why is it that a sperm whale can be 500 times our size but its tapeworms are not much larger? One reason is that humans (at least in much of the developed world) can always find more food to eat if they are hungry. The record beef tapeworm probably took 25 years or so to grow to 72 ft (22 m), and for much of that time it would have been consuming a large fraction of the food its female host swallowed. However, the host was likely able to compensate for

▲ The giant earthworm (*Megascolides australis*) reaches lengths of 10 ft (3 m). This researcher has tied a knot in the worm to stop it from pulling itself back into its burrow.

◀ The beef tapeworm (*Taenia saginata*) grows to 72 ft (22 m) in length. To avoid infection with this intestinal parasite, avoid eating undercooked meat.

▶ A tapeworm inside a human intestine, along with a close-up of the head, which the parasite uses to attach itself securely to the wall of the digestive tract.

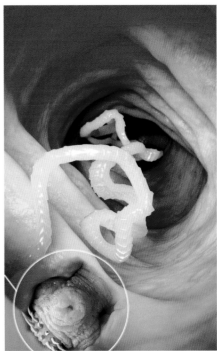

Bootlace worm
Lineus longissimus
Length: possibly up to
55 m (180 ft)

Given how they twist and
coil around themselves (see
opposite), it is near impossible
to measure the length of a
long living bootlace worm.
With dead specimens, it is
hard to be sure how much
they have changed post
mortem in ways that
influence their length.

this simply by eating much more than she normally would have done without the parasite inside her. In contrast, this option is not open to wild animals that are already working quite hard to find the food they need. If a tapeworm took up a large fraction of the digestive tract of a wild animal, then the host would fail to meet its metabolic requirements, causing it to waste away and hence become less able to forage for itself, in turn causing it to waste away even quicker until eventually it dies. In that sense, modern humans in the developed world are an ideal host for a tapeworm, but in another sense, we are a poor choice because modern medicine can normally help us shed tapeworms and other internal parasites.

The Longest Organism?

Earlier in this chapter, I credited the lion's mane jellyfish as probably the longest animal alive. Our uncertainty stems from the report of a 180 ft-long (55 m) bootlace worm (*Lineus longissimus*), a species that is common along the coasts of Britain and Norway. No live individual larger than 30 ft (10 m) has ever been recorded, but in 1864 a dead individual washed ashore at St Andrews in Scotland, a five-minute drive from where I am typing this. The worm doesn't seem to have been preserved, but a contemporary report says that it "half filled a dissecting jar eight inches wide and five inches deep. Thirty yards were measured without rupture, and yet the mass was not half uncoiled". If the measured "not half uncoiled" portion was 30 yd (27.43 m) long, then

the total length of the worm must have been at least 180 ft (55 m), so that is the value commonly reported in popular science books and on websites. However, bootlace worms stretch considerably when they are being buffeted about by waves after death, so it is very hard to know the true size this animal would have been in life.

I have seen calculations that focus on the other half of the quote above, asking how long the worm would have been to half-fill the jar given a range of plausible assumptions about its width, and again 180 ft (55 m) seems plausible. However, we don't know the state of decay of the animal, or whether bootlace worms take on water after death or produce gases during decomposition that would have caused the individual to swell. In addition, this short quote doesn't say whether the worm half-filled the jar without any airspaces or was loosely coiled, so we are guessing pretty wildly on the basis of very flimsy evidence. It is certainly possible that this species includes individuals that are the longest organisms alive, but it would be nice to have a bit more evidence. In fact, the evidence for the long length of the lion's mane jellyfish (see page 156) is just as flimsy, stemming from an 1865 report that says, "I measured myself a specimen... The tentacles extended to a length of more than a hundred and 20 feet." All the caveats I mention above also apply to this claim, so we really have no definitive idea which species is the longest, and *Polygonoporus giganticus*—for which we do have good documentary evidence—might not be out of the competition yet.

▶ Few animals are as aptly named as the bootlace worm—although it could more accurately be called the knotted bootlace worm!

Chapter 8

RECORD REPTILES AND AMPHIBIANS

The first vertebrates to emerge from the oceans were the amphibians around 370 million years ago, and in terms of big animals, this group had the land to themselves for more than 100 million years. Amphibians have water-permeable eggs, and so must lay these in wet or damp conditions to prevent the embryo drying out. The key innovation of the reptiles, which rose to dominance afterwards, was a tough shell to the eggs, allowing them to be laid anywhere. Here, we explore the biggest of these two groups of vertebrates.

Scaled-up Snakes

The worms discussed in Chapter 7 are invertebrates, lacking the spine of humans and other vertebrates. Instead, they have a hydrostatic skeleton whose rigidity comes from internal pressure, much like a bicycle tire. While snakes may resemble worms superficially, being long and thin, they actually have an entirely different body plan. They are vertebrates like us, but whereas our spine comprises thirty-three vertebrae, a snake might have 300, making it far more flexible.

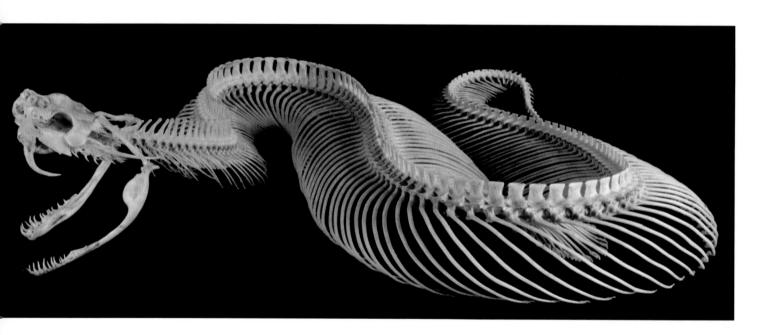

Squeezing and Biting Snakes

Many humans have a healthy fear of snakes—indeed, it is said to be the most commonly reported phobia. This makes perfect sense, as lots of snakes are downright dangerous and plenty can kill with relative ease. There are two quite different ways in which snakes kill their prey: through a venomous bite; or by coiling themselves around the animal and squeezing it so tight that it can't expand its ribcage to breathe, and possibly even stopping the blood flow to its brain. In truth, we have much more to fear from the biters than the squeezers: it is estimated that around 90,000 people are killed by venomous snakebites each year, compared to less than nine a year that are squeezed to death. Perhaps not surprisingly, the biggest snakes are squeezers (or constrictors, to use the more conventional term)—while a venomous snake doesn't need to be big to kill a human, a constrictor certainly does.

The reticulated python
(Python reticulatus)
Length: 20 ft (6.1 m)

The reticulated python is one of the world's longest snakes, with some individuals exceeding 20 ft. Well-documented cases from Indonesia in 2017 and 2018 prove that it can kill and eat an adult human whole.

▲ Snakes are vertebrates, just like us, with an extensive skeleton to support the muscular body that allows many of them to squeeze the life out of their prey.

Bone Crushers

There are some very big constricting snakes indeed. The green anaconda (*Eunectes murinus*) can reach lengths in excess of 16 ft (5 m) and weigh almost 220 lb (100 kg); a Southeast Asian reticulated python (*Python reticulatus*) might be even longer, at 23 ft (7 m), but not quite as heavy, at 130 lb (60 kg); and the Burmese python (*Python bivittatus*), also found in Southeast Asia, can grow to 16 ft (5 m) and 165 lb (75 kg). All of these species have been implicated in the deaths of humans by constriction. Evidence for such attacks is anecdotal, since these species live in sparsely populated regions. However, both of the large pythons have been popular as pets, and surprisingly the country with the highest death rate caused by snake constriction may well be the USA, when python pets grow too large to be controlled by their owners.

Having squeezed the life out of its victim, a constricting snake will then generally attempt to eat it whole. The three large constrictors can certainly swallow large prey, and the biggest individuals could consume an adult human. As a general rule of thumb, a snake can swallow a prey item weighing 60 percent of its own body weight easily, and meal sizes up to 100 percent of body weight are not unheard of. Although the slimmer reticulated python might struggle with a man's shoulders, there are reports of people being eaten by this and the other colossal constrictors, including cases in Indonesia in 2017 and 2018.

▼ In 2017, a twenty-five-year-old Indonesian farmer went missing. During the search for him, a reticulated python with a huge bulge in the middle of its 23 ft (7 m) body was found, which unfortunately turned out to be the farmer. In 2018, a fifty-four-year-old woman working in her garden was similarly the victim of a reticulated python.

Green anaconda
Eunectes murinus
Length: 20 ft (6.1 m)

The longest individuals of this species are similar to the longest pythons, at more than 20 ft, but the anaconda is thicker and therefore heavier—the largest individuals weigh at least 350 lb (160 kg).

Burmese python
Python bivittatus
Length: 16 ft (4.9 m)

Only a little smaller on average than the reticulating python, the Burmese python (*Python bivittatus*) is surprisingly popular as a pet. However, pets can escape or are released, and there is now a thriving population of these snakes in the Florida Everglades.

Titanic Boa

It is easy to understand why constrictors might evolve to become large. In the absence of venom, battles between males for access to females are trials of strength, with the winner often being the larger of the pair. Snakes in general have to endure long periods between meals, and as body size increases, so the fat that can be stored to see the animals through lean times should increase more quickly than metabolic rate, meaning that larger individuals are less at risk from starvation. Additionally, the bigger constrictors are, the bigger prey they can tackle.

Titanoboa cerrejonensis
Length: 50 ft (15 m)

Titanoboa cerrejonensis is the largest species of snake known to have existed, reaching perhaps 50 ft in length and weighing 4,400 lb (2,000 kg). Very few animals living in South America alongside these snakes 60 million years ago would have been large enough to be safe from their jaws.

The biggest snake ever found is *Titanoboa cerrejonensis*, fossils of which indicate it lived around 60 million years ago in what is now South America. It was similar to an anaconda, but might well have been more than 43 ft (13 m) long and 1,750 lb (800 kg) in weight. The strange thing about *T. cerrejonensis* is that individuals seem to have reached a huge size with regularity. Paleontologists have found remains of eight separate individuals, all of which seem to be equally huge, with an estimated length of 36–50 ft (11–15 m). Vertebrates generally show considerable variation in body size, although extremes are highly unusual. In *T. cerrejonensis*, however, it seems that the snakes grew readily to some limiting size—although scientists can't be sure of this until more remains are found. My guess is that the ancestor of this snake was nowhere near as big, as it is hard to imagine such a huge top predator making it through the aftermath of the cataclysmic meteor strike. Rather, it may have grown big opportunistically in a world now devoid of dinosaurs but where mammals had yet to diversify to fill the top predator roles.

Deadliest Snakes

The inland taipan (*Oxyuranus microlepidotus*) of central Australia is said to be the world's most venomous snake, and a single bite can transfer enough toxin to kill a hundred adult humans. Worse still, unlike most snakes the species doesn't seem to deliver dry "warning" bites and will almost certainly transfer venom if it does strike, and also unlike most snakes, it often bites several times. Left untreated, victims can be dead within the hour. Despite this, there are very few records of people being bitten. This is because the inland taipan is rare, it lives in locations only sparsely inhabited by people, and it has a shy and retiring nature.

The African black mamba (*Dendroaspis polylepis*) is responsible for far more deaths. Like the inland taipan, it has a particularly potent venom, it doesn't deliver dry bites, and it often strikes multiple times. However, in contrast with the Australian species it has a large natural distribution, occurs in a wide diversity of habitat types, and is considered to be quite numerous. In addition, a lot of people live within its range, and the snake tends to be territorial and is fast-moving. It is likely to start rearing up and displaying if you get within 100 ft (30 m) of it, and if you don't heed this warning then there is a very real possibility you will be attacked.

▲ Black mambas (*Dendroaspis polylepis*) are among the most feared venomous snakes, but this does not deter mongooses from preying on them. The mongooses benefit from speed, teamwork, and partial immunity to the snake's venom.

Within Limits

It is not quite clear why there are no snakes as monstrous as *Titanoboa cerrejonensis* today. It might be that we live in unusual times where large animals are not particularly common. Thus, green anacondas may simply be as big as they need to be—in other words, they don't really encounter anything that is too large for them to eat, so are not under any pressure to grow larger themselves in order to expand their dietary choices. The scientific paper that first reported on the *T. cerrejonensis* find in 2009 suggested that the snake lived a world considerably warmer than our modern one, and that these higher temperatures were essential for a huge cold-blooded ambush predator to stay warm enough to function properly. I am not so sure, however, as modern-day saltwater crocodiles (see page 188) function very well as cold-blooded ambush predators, and they weigh up to 2,200 lb (1,000 kg)—at least as heavy as *T. cerrejonensis*. We will look at this again later in the chapter (see page 180).

▲ *Titanoboa cerrejonensis* lived in swampland, and crocodiles may well have featured in its diet, although the large turtles that apparently also shared its habitat would have been an easier meal.

Dragons and Monitors

Komodo dragons (*Varanus komodoensis*) are the largest living lizards, but here we also touch on some other lizard leviathans, one of which might be even longer than the Komodo and another that is much easier to find in the wild. Our tour of the giant lizards ends with a focus on an Australian ancestor.

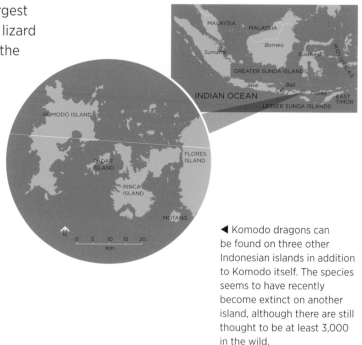

The Island Dragon

The Komodo dragon is the largest lizard today, reaching a length of 10 ft (3 m) and weighing upwards of 150 lb (70 kg). The species is found on a scattering of Indonesian islands, including Komodo itself, and it seems likely that giants like this once existed widely across Asia and Australia but were outcompeted by mammals. The Komodo has a varied diet, being both a predator and a scavenger, but most of its food comes from killing deer. The lizards generally avoid human contact but they have attacked people—and some of

◄ Komodo dragons can be found on three other Indonesian islands in addition to Komodo itself. The species seems to have recently become extinct on another island, although there are still thought to be at least 3,000 in the wild.

Flores—An Island of Giants and Halflings

One of the Indonesian islands with a wild population of Komodo dragons is Flores, which neighbors Komodo Island. Flores has several examples of small species that have become bigger than elsewhere, like the Komodo dragon, and large species becoming smaller (see Chapter 1). One of the most extraordinary examples of the latter was discovered in 2003, when the fossil remains of small human-like individuals dating back 50,000 years were found in a limestone cave on the island. The hominins had a maximum adult height of a little under 4 ft (1.2 m), so it's no surprise that they have been nicknamed hobbits. Scientists haven't agreed yet on why they were so small. Some see them as a population of our own species, *Homo sapiens*, that suffered from some pathological condition that stunted growth. Others view them as a separate species (called *Homo floresiensis*) that derived from an earlier species in our ancestry (*Homo erectus*), which arrived on the island as far back as a million years ago and subsequently evolved to be smaller.

Another fascinating question is how the hobbits got there. There have been sea-level changes over the last million years, and at times it would have been possible to walk from Flores to Komodo, but not from Komodo to the mainland. Some scientists believe that the existence

of human-like folk on Flores argues for early boat building (and thus cooperation and language). However, I don't think we should rule out the possibility that a small number of individuals were washed there entirely inadvertently, perhaps following a giant tsunami. This is an exciting new twist in our ancestry, and hopefully paleontologists will find more fossil remains on Flores (and even Komodo) to help us piece the story together.

▼ *Homo floresiensis* individuals weren't just shorter than modern humans—their head, and hence their brain, was also a lot smaller too.

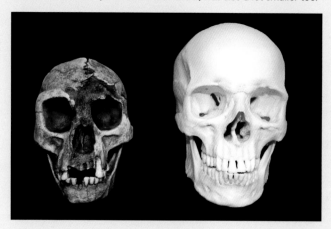

these attacks have been fatal. Part of the hazard lies in the fact that the lizard has a venomous bite that prevents blood from clotting. They are also known to prey on human corpses and have an excellent sense of smell, so burials are particularly deep on islands where the Komodo occurs, and rocks are piled on top of graves. The feeding behavior of the dragons additionally extends to attacks on smaller members of its own kind, and Komodo youngsters spend much of their time in trees to avoid such a fate.

Its impressive size makes the Komodo dragon a popular zoo exhibit, and this has recently led to the discovery of a very unusual aspect to the species' biology: a female kept on her own can still produce

viable eggs despite having never bred with a male. Offspring from such eggs are always male. This type of behavior is known as a lifeboat mechanism, and is a strategy used to save a population on the edge of extinction. Imagine that a population has dwindled to a single individual. If that individual is male, then there is no hope the population will recover without introductions from elsewhere. But if that lone individual is female, then potentially she could give birth to males and then breed with one of them to produce more male and female individuals, thereby allowing the population to recover. This strategy is not foolproof, however—on the island Padar, the Komodo dragon seems to have gone extinct in the last few years, and the populations on some other smaller islands are also declining. In all these cases, the apparent cause of the decline is human overhunting of deer and other large herbivores, depriving the dragons of their food source. For the time being, the populations on the biggest islands—Komodo and Rinca—seem stable. The perceived value of the species to tourism and the human-shunning instincts of most individuals should help to protect it going forward.

Komodo dragon
Varanus komodoensis
Length: 10 ft (3 m)

The Komodo dragon) is the largest living lizard, growing to more than 10 ft (3 m) long and weighing up to 150 lb (70 kg). Amazingly, this giant was unknown to science until just before the First World War.

Crocodile monitor
Varanus salvadorii
Length: 8 ft (2.4 m)

The crocodile monitor can be almost 8 ft long, but it is very thin, with the largest individuals weighing perhaps less than 44 lb (20 kg). This explains why these animals are frequently seen in trees.

A Whip-tailed Giant

The crocodile monitor (*Varanus salvadorii*) from New Guinea is sometimes cited as the world's longest living lizard. It is much more slightly built than the Komodo dragon, and even large adults spend a lot of time up trees, but it has a very long tail that in some individuals is perhaps twice as long as the rest of the body. The tail is likely useful for balance in trees and is also used in a whipping motion in self-defense.

What makes the crocodile monitor an unusual lizard is that it has an active hunting style rather than the ambush predation seen in many large reptilian predators. The key adaptation that makes this possible is its ability to run and breathe at the same time—something that we take for granted, but that is denied to almost all lizards. Most reptiles have a running style that involves flexing their body from side to side, but this causes a problem in that the lung on the side that is compressed does not have enough space to inflate. This doesn't prevent lizards from running fast, but it does prevent them from running for a long time. In the case of the crocodile monitor, however, its throat has adapted to overcome the problem, acting as a kind of pump to draw in more air, thereby allowing the lizard to chase down prey. Happily, the species generally avoids human contact.

Swamp Monster

The easiest giant lizard to see in the wild is the water monitor (*Varanus salvator*), which unlike the Komodo dragon and crocodile monitor has a wide distribution across Asia, is quite common, and is not especially shy of people, to the extent that it even lives in relatively urban environments. As its name suggests, it spends a lot of time in water, so if you visit South or Southeast Asia, it's worth having a look for one in any canal or river you happen to pass. The water monitor is a dark, muscularly built lizard that can be up to 6 ft (2 m) long and weigh more than 110 lb (50 kg), making it second only to the Komodo dragon in terms of weight. Not many individuals reach that size, however, since more than a million are hunted each year for their skins (which are turned into shoes, belts, and handbags), for the live pet trade, for food, and for traditional medicine. One of the best places to see them is Sri Lanka, where they are widespread but hunting is uncommon due to the perception that although monitors might steal the odd chicken, they do more good by controlling the numbers of venomous snakes and other pests.

Water monitor
Varanus salvator
Length: 6 ft (2.5 m)

The water monitor can exceed 6 ft in length, with the heaviest individuals weighing nearly as much as an adult human. The lizards are often quite at home in environments disturbed by humans and even in urban areas—in fact, they are easy to spot in canals and parks in Bangkok, Thailand.

Ancient Great Roamer

For a period of a million years or so until about 40,000 years ago, a large lizard similar to the living giants described above roamed southern Australia. Scientists can't quite agree on its relatedness to other lizards, so it is called either *Megalania prisca* or *Varanus priscus*. Because very few complete or near-complete skeletons have been found, it is difficult to know how big it actually was. However, it is estimated that the largest individuals were around 20 ft (6.1 m) long (twice as long as a Komodo dragon) and weighed more than 1,100 lb (500 kg). This would have made the species an apex predator—at the top of the food chain, and able to prey on almost any other species and itself vulnerable to few. Its main rival would have been the 350 lb (160 kg) marsupial lion (*Thylacoleo carnifex*), which seems to have been much more commonplace judging by the relative numbers of remains found. Both species became extinct around the same time, which more or less coincides with the arrival of humans in Australia. It is very tempting to link these events, but in truth the early human habitation of Australia is far from understood.

One interesting point is that, since these two species died out, Australia has been conspicuously short of a large terrestrial apex predator. Only the 65 lb (30 kg) thylacine (*Thylacinus cynocephalus*) remained, and this disappeared on the mainland a few thousand years ago and lingered on in Tasmania until it died out sometime in the twentieth century. In this case, hunting and habitat change (through fire) by humans, and perhaps also competition from the dingoes (*Canis familiaris*) they introduced, are potential culprits. Australia was the last large landmass to be populated by humans, who were already technologically sophisticated when they arrived. Apparently, humans were such efficient top predators from the get-go in Australia that they left little room for other large predators.

◄ Although often found in close proximity to humans, water monitors only very rarely attack people or pets. Being highly adept swimmers, they mostly eat fish and frogs.

Varanus priscus
Length: 20 ft (6.1 m)

Varanus priscus might have been 20 ft long and weighed more than 1,100 lb (500 kg), and it seems to have still been alive in Australia when humans first arrived on the continent.

Finding Your Niche

The 66 million years since the demise of the dinosaurs have seen the rise of mammals and birds as the dominant large animals across most terrestrial parts of the world, pushing reptiles and amphibians to the edges. When asking why some large reptiles still do occur on land, it is important to consider just how much mammals and birds need to eat.

◀ Tortoises (like this giant from the Galápagos Islands) are herbivores, and are often found in environments that aren't suited to mammalian herbivores like rabbits, goats, and deer.

Food for Thought

Why the Komodo dragon and other large reptiles have thrived is because they fit into (different) niches that would not work for a mammal. As we saw in Chapter 3, mammals, like birds, are endotherms—in other words, they maintain a high and relatively constant body temperature by means of internal heat production. When you consider that the metabolism of a mammal at rest can be ten times that of an equivalent-sized reptile, it becomes clear that this is a very expensive way to live. Accordingly, mammals need to eat a lot and big mammals in particular need to eat a huge amount, so a viable population of these animals needs

a substantial food supply. This can be a problem on isolated islands, where there is only so much food to go around, and especially acute for large predators at the top of a food chain. Imagine a simple food chain where rabbits eat grass and in turn are eaten by foxes. The problem for the foxes is that when they eat the rabbits they don't get all the energy the rabbits got from the grass. Instead, they just get the energy the rabbits spent on growing and putting on fat and muscle, not the much larger amounts of energy the rabbits used to fuel their own metabolism. So, the amount of energy available to a population of animals declines the higher up the food chain it feeds.

On Komodo Island, there is a large enough population of deer to support a viable population of Komodo dragons, but not enough for the same population size of, say, tigers. This is because tigers are mammals and so need to eat much more. In fact, the tiger population that could be supported might be only a tenth the number of individuals of the dragon population. Such a tiny population would be very prone to extinction—if, for example, one animal becomes sick, another is injured, and another fails to breed, suddenly numbers have dwindled to nothing. For the reason that the food base is smaller on islands, we therefore generally find reptiles there filling niches that on the mainland are occupied by mammals. Hence, carnivorous Komodo dragons occur on small islands in Indonesia rather than tigers or wolves, and giant tortoises are found on islands around the world filling niches often taken up by rabbits and deer on the mainland.

Feast or Famine

Like the Komodo dragons and giant tortoises mentioned above, crocodilians also fill a niche that a similar-sized mammal would struggle with. The attraction of being an ambush predator is that you save a lot of energy that would otherwise be spent looking for prey by simply waiting for it to come to you. The drawback to this is that you might have to wait a long time between meals. This is not a problem for a big crocodile or snake, however, which have enough reserves to last for up to six months between meals. In comparison, big cats or bears need to eat regularly and would certainly start to feel weak with hunger after a week without food.

We have all seen nature documentaries on TV showing footage of Nile crocodiles (*Crocodylus niloticus*) feasting on zebras and wildebeest fording rivers during their annual migration. Such food bonanzas occur only once or twice a year, but the crocodiles can survive (and even thrive) on these very occasional feasts, supplemented by the odd dead bird or fish that floats or swims past. As with other reptiles, it is the low food requirement of the crocodiles that allows them to fill a niche that would not work for a mammal.

▲ Crocodiles spend only a small fraction of their time feeding; mostly, they doze or keep cool by opening their jaws wide, like this one.

◄ Just one impala (*Aepyceros melampus*) would sustain even a large crocodile for several weeks.

Giant Tortoises

We don't really think of tortoises as the most mobile of creatures, but they have been surprisingly adept colonizers of even remote islands. Part of the reason for this is that they float well thanks to air trapped under their shell, but they are also often good swimmers and can last a long time without food or fresh water. On remote islands they often grow really big thanks to a lack of competition from mammalian herbivores like deer.

Slow and Steady

Tortoises have been around for 250 million years or more, and there have been giant tortoises for at least 70 million years. Once, giant tortoises were widespread on continental landmasses as well as on many islands, but today they are confined to just two island groups. What really did for many island tortoises were European sailors in the sixteenth to nineteenth centuries, simply because large individuals offer a lot of meat. It was only

natural for sailors to break their journeys on suitable islands to allow them to stretch their legs and stock up on useful materials like wood, as well as fresh food and water. Giant tortoises were found to be an ideal food source, being easy to spot and too slow to run away, and especially because they could survive in the hold of a ship for several months without food or water. This made them a great source of fresh meat long after fruit had rotted and other live animals gathered at the same time had died.

The Indian Ocean islands of Mauritius, Réunion, and Rodrigues were home to five species of giant tortoises until they were wiped out by humans over the last 400 years, so it is particularly pleasing that giant tortoises have now been reintroduced to two of these islands (see below). In fairness, humans aren't entirely to blame for the extinction of species of giant tortoise. The Canary Islands once boasted two species of giants, both of which appear to have died out before there was any human contact with the islands— probably when volcanic eruptions destroyed so much of their habitat that it could no longer provide them with sufficient food.

Giant tortoises still found in the wild today include a population of 150,000 Aldabra giant tortoises (*Aldabrachelys gigantea*) on the remote Aldabra Atoll in the Indian Ocean, a small population on the island of Changuu near Zanzibar, and some thriving, recently introduced populations on Mauritius and Rodrigues. In addition, there are 2,500 Galápagos tortoises (*Chelonoidis* species) on various of the Galápagos Islands in the Pacific Ocean. In all of these remaining species, the longest-living individuals can grow to just over 880 lb (400 kg).

It is relatively easy to see how such giants reached remote islands: tortoises generally float very well thanks to air trapped under their shell, and as discussed above, they can survive for long periods without eating or drinking. It is not impossible to imagine a group of mainland giant tortoises washed out to sea by a storm or tsunami, and then floating on the ocean currents and making landfall on the same island. While such an event is unlikely, it is not impossible. After all, there are several large tsunamis around the world each century and giant tortoises have been around for tens of millions of years.

Galápagos giant tortoise

Chelonoidis sp.
Weight: 880 lb (400 kg)

There are ten different species of Galápagos giant tortoise remaining today. All are pretty similar in size. It is rather painful to reflect that there were fifteen different species when the naturalist Charles Darwin visited the islands in 1835, and that he pronounced that "the young tortoises make excellent soup."

George and Jonathan

Lonesome George was the last known Galápagos Pinta Island tortoise (*Chelonoidis abingdonii*). By 1971, the vegetation on the island on Pinta had suffered terribly from browsing by introduced feral goats, and the tortoise population had crashed. A single male was discovered and taken into captivity while a search of the island and of zoo collections was made in the hope of locating a female. None was found, however, and Lonesome George, as he was called, became a conservation icon. He died in 2012, and with him, his species was gone forever.

Jonathan is a Seychelles giant tortoise (*Aldabrachelys gigantea hololissa*) and is generally considered to be the oldest living terrestrial animal. In 1882, he was brought by boat to the small island of St. Helena in the South Atlantic and presented as a gift to the British governor of the island. Astonishingly, he is still alive today. We don't know Jonathan's exact age, but a photograph taken of him in 1886 indicates that at the time he was fully mature in both size and appearance—a state not generally reached in giant tortoises until the age fifty. So, if Jonathan was fifty in 1886, then that made him at least 182 years old in 2018.

▼ Below: Jonathan is a Seychelles giant tortoise (*Aldabrachelys gigantea hololissa*), although he hasn't seen his home islands since he was removed from them in 1882. Latterly, he hasn't seen anything at all, as he seems to have been blind for some years.

▼ Bottom: Lonesome George remains a conservation icon. Following his death in 2012, he was preserved taxidermically and is now on display in the Charles Darwin Research Station in the Galápagos.

Giant Sea Turtles

Leatherback turtle
Dermochelys coriacea
Weight: 1,400 lb (600 kg)

A leatherback turtle feeding on a sea pickle, a large marine invertebrate in the genus *Pyrosoma*. Although their diet is not especially nutritious, the turtles spend a lot of their time feeding and can grow to weigh more than half a ton.

While tortoises have evolved to giant sizes on some islands (see page 182), living in water can offer animals the freedom to reach even larger sizes. No land tortoise can compare with the size of leatherback turtles (*Dermochelys coriacea*), which spend almost their whole lives at sea. In fact, the leatherback is the fourth-largest living reptile of all, beaten only by three species of (again, mostly aquatic) crocodiles.

There are several species of sea turtle, all fairly large, but easily the biggest is the leatherback turtle. It is almost fully aquatic, needing sandy beaches only to lay its eggs, and otherwise swims the oceans in search of jellyfish prey. The largest individuals can reach 1,400 lb (650 kg).

Keeping Warm

Leatherbacks are found all around the world, even in surprisingly cold waters like those off Alaska and Norway. Their search for food also takes them to depths below 3,000 ft (1,000 m), where the waters are always cold. They manage to cope with these cold conditions because, unlike most other reptiles, they are endothermic. They generate a lot of heat in their muscles by swimming almost constantly, and they retain this heat very well thanks to a number of factors. For a start, their big size helps—as discussed in Chapter 1, the larger an object, the greater its volume compared to its surface area. In the case of the leatherback, it has a high volume of muscle compared to the surface area from which it might lose internally generated heat. In addition, it has a thick covering of fat around its body, just like the blubber of whales, which acts as insulation.

In order to swim efficiently, leatherbacks have huge front flippers that are easily 8 ft (2.5 m) long in the largest individuals. They are also relatively flat and so have a high surface area through which heat could be lost—a particular problem, as the turtle has to send blood to the flippers for normal metabolic functioning of the tissues there. The animal has managed to overcome this apparent drawback, however, by having a clever counter-current heat exchanger that strips heat out of blood flowing into the flippers and uses it to rewarm blood coming back into the core of the animal from the flippers, thereby considerably reducing heat loss. In this way, the leatherback can keep the core of its body as much as 32°F (18°C) warmer than the waters through which it is swimming. This high-intensity mode of life means that the turtle has to feed almost all the time, and because its diet is made up almost entirely of jellyfish, it also needs to swim almost all the time in search of its prey. Despite their low food value (they are about 95 percent water), jellyfish are an attractive food source as they are plentiful in the oceans and the turtles don't have to swim fast to catch them.

▼ Heaving themselves across the sand must be a huge effort for female leatherbacks, which is why they prefer beaches that aren't too steep and where they don't have to crawl far to reach sand above the high-tide line.

▲ Most turtle hatchlings, like these leatherbacks, emerge from the nest together as a way of swamping local predators while they all make their dash down the beach to the sea. Once in the sea, the males will spend the rest of their lives there, never returning to land again.

A Numbers Game

Being large, adult leatherbacks are reputedly able to hold predatory sharks and even orcas (*Orcinus orca*) at bay, so they have few predators to worry about. However, they are very vulnerable at the egg stage and as young hatchlings. The eggs need to incubate for two months in warm sands in order for the offspring to reach the hatchling stage, and during that time all manner of mammals, birds, crabs, and monitor lizards have been observed digging them out for a bountiful and nutritious meal. The female leatherback doesn't guard her offspring, instead returning to the ocean as soon as she has laid the eggs, and in any case, she is very ungainly on land. Worse still, the beaches used for egg-laying are pretty predictable—they must have sand that is warm, free from stones (since the underside of the adults is not well protected), and soft enough to allow digging, and a shallow gradient since the female would struggle to propel herself up a steep slope. Beaches that meet all these criteria are not especially common, so predators can focus their attentions effectively, and mortality among eggs and hatchlings is consequently very high.

Even when they reach the sea, the hatchlings are still vulnerable to marine predators for several years, leaving only a very lucky few to attain large size. The biggest threat to adult leatherbacks is not natural predators but plastic bags. Disposable grocery bags can easily be blown into the sea if discarded carelessly, and they then float about in the surface waters for years. Unfortunately for leatherbacks, plastic bags look enough like jellyfish to fool them, and ingested plastic can jam up the innards of a turtle entirely, leading to a slow death by starvation.

Extinct Shelled Giants

The last 66 million years have seen the progressive rise of mammals to fill many ecological niches, especially those requiring a large body size. This means that when we look at the fossil record, we tend to find that giant tortoises and turtles were once more diverse and more widespread than they are today. Many niches that were once filled by tortoises are now filled by rabbits and deer, and many of those that were once filled by giant turtles are filled by marine mammals.

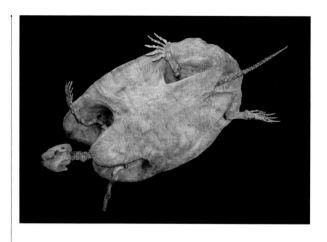

Stupendemys geographicus
Shell length: 8 ft (2.5 m)

Stupendemys geographicus lived in South America 6–5 million years ago and is one of the largest turtles known. Its remains have been found in sediments that suggest it lived in fresh water rather than the sea. It likely wasn't the strongest swimmer, so was perhaps a swamp dweller during the Late Miocene–Early Pliocene epoch.

Shell Shock

As is the case with many groups of animals, there were larger tortoises and turtles in the past than there are around today. The largest tortoise found to date is a species called *Megalochelys atlas*, which seems to have been quite widespread across Asia between about 10 million and 2 million years ago, and was nearly 6 ft (2 m) tall and 10 ft (3 m) long. Estimating the weight of such a monster is very tricky, since relatively small changes in the assumed thickness of the shell and how

completely it was filled with flesh substantially affect the estimates, but figures start at 2,200 lb (1,000 kg) and go all the way up to 8,800 lb (4,000 kg).

Meiolania was a genus of tortoises that reached almost the same size as *Megalochelys atlas*; these lived during the same period, but in Australia and some Pacific islands. They were unusual in that neither the head nor the tail could retract fully into the shell for protection, and as a result both ends were heavily armored and the animals even had large horns. These beasts survived on the remote Pacific archipelago of New Caledonia until as recently as 3,000 years ago, and the grizzly remains of bones found in ancient garbage middens indicate that hunting likely drove them to extinction within the first few years of human habitation of the islands.

Stupendemys

Archelon

Megalochelys

0   1   2   3   4
Meters

***Meiolania* sp.**
Shell length: 8 ft (2.5 m)

Members of the genus *Meiolania* lived on land in Australia 50,000 years ago and had a shell measuring around 8 ft in length. The head and tail weren't pulled into its shell when the animal was threatened, but instead were heavily armored, with the addition of horn-like projections on the head.

Roaming the oceans about 80 million years ago were mammoth marine turtles in the genus *Archelon*, estimated to weigh at least 4,800 lb (2,200 kg) when fully grown. These had much more robust mouthparts than modern-day leatherbacks and appear to have had a very powerful bite, so they probably had a crunchier diet than the jellyfish of the leatherback, perhaps exploiting swimming mollusks like squid. An even bigger freshwater turtle, called *Stupendemys geographicus*, lived around 6 million or 5 million years ago in the slow-moving rivers and swamplands of South America. It seems to have been a relatively weak swimmer judging by the likely weight of its shell and estimates of the size of its limbs and the muscles available to move them. The turtle's great mass and relatively flat shape might have been an adaptation to river currents, allowing it to stand firm on the riverbed while it grazed. Its large bulk would also have enabled it to stay submerged for longer.

***Archelon* sp.**
Length: 16 ft (5 m)

Archelon species were huge oceanic turtles that lived 80 million years ago. Their closest living relative is the leatherback (*Dermochelys coriacea*), and they had a similar carapace that mostly comprised thickened hide rather than bone.

Surviving an Extinction Event

As we saw in Chapter 4, there were many different types of reptiles in the oceans and fresh waters at the time dinosaurs ruled the land, but almost all were wiped out with the dinosaurs when a meteor hit the Earth 66 million years ago. However, two lineages of aquatic reptiles did survive the extinction event and are still with us today: turtles and crocodiles. So, why did they survive when so many others did not?

My guess is that it is no coincidence that most turtles and crocodiles live in freshwater rivers and not in the seas, because the ecosystem type least affected by the Cretaceous–Paleogene extinction event was river habitats. Riverine species were still terribly badly affected by the catastrophe and many were driven to extinction, but not quite as many as in other types of habitat. This might be because an important foodstuff in river systems is detritus—material cast off by other organisms, whole bodies of dead organisms that fall to the river bottom, and dead vegetation that falls or is blown in. It can take years for this material to break down, and a thick layer of it accumulates on the riverbed, providing an important source of food for small riverine creatures.

When the meteor hit, great clouds of dust covered the Earth and shut down photosynthesis, killing plants and the animals that depended on them. However, animals living in freshwater ecosystems would have been buffered by the supply of food in the detritus layer, helping them—and animals that ate them—to survive while the atmosphere cleared and the Earth's vegetation slowly recovered. I suspect that a few crocodiles and turtles survived in river systems in this way, and then reinvaded the seas from these strongholds.

Stunning Salties

The three biggest living reptiles are all crocodiles—the saltwater crocodile (*Crocodylus porosus*), the Nile crocodile (*C. niloticus*), and the Orinoco crocodile (*C. intermedius*)—and all are large enough to be man-eaters. We will focus on the salties first, as the biggest of them all. Their size, combined with an aggressive territoriality, means they are a real risk to the unwary swimmer and even to those in boats.

Saltie Statistics

The saltwater crocodile, commonly nicknamed the saltie, is definitely the largest of the living reptiles, with record males exceeding 20 ft (6.1 m) in length and weighing 2,600 lb (1,200 kg). It is also the most widespread of the crocodile species, ranging from India through most of Southeast Asia to the northern coast of Australia. As its name suggests, it is found in saltwater, although the species seems to use the open ocean only for commuting longer distances and generally lives in river deltas and estuaries. It's the most aquatic of all the crocodiles: many other species spend much of their day dozing on riverbanks and will even feed out of the water, but salties can spend weeks at a time without coming ashore. To round up this impressive list of statistics, salties are considered to have the most powerful bite of any animal (see box opposite). Males are fiercely territorial, and will often attack anything that dares to float into their stretch of river—reports of unprovoked attacks on boats and even on tree trunks floating downriver are commonplace. This aggression, coupled with the immense size and power of these reptiles, means they can be dangerous to humans. The need to compete for territory explains why males are generally substantially larger than females.

The saltwater crocodile
Crocodylus porosus
Length: 20 ft (6.1 m)

Although the saltwater crocodile is unusual in being comfortable in saltwater, the species is a typical crocodile in that it is relatively sluggish for much of the time. That said, if presented with the chance to ambush prey, it can summon up the power to launch its huge bulk out of the water in an instant.

Bite Force

Another claim to fame of the saltwater crocodile is that it has the greatest bite force of any animal. In a laboratory study, a 17 ft-long (5.2 m) individual had its bite force measured at more than 3,600 lbf (16,000 N). To put that in perspective, if I stood on your back while you were lying face down, gravity acting on my mass would impose 180 lbf (800 N) on you, so the bite force of that saltie was like having twenty fully grown men standing on you. There is not a bone in your body the croc could not have snapped like a twig. The scientist who took the original measurement then applied the physics of levers to estimate how much force a 20 ft (6.1 m) saltie could generate, and came up with a figure of 7,600 lbf (34,000 N), equivalent to the weight of forty-two adult males. He carried his calculations a bit further, reckoning that a 40 ft-long (12 m) crocodile like the extinct *Sarcosuchus* species (see page 194) would have a bite force of 23,100 lbf (103,000 N), or as much as the weight of 130 grown men and twice the estimated bite force of a *Tyrannosaurus rex*.

Crocodiles generate this force by having particularly huge jaw muscles, which are visible as bulges at the side of the head where you would expect their jaws to hinge.

In addition, they devote almost all of that muscle power to closing their jaws and very little to opening them. So, if you are ever unlucky enough to be grappling with a crocodile and its jaws close, remember that you are probably strong enough to hold them closed!

▼ Bulges at the side of the head of a saltie hold the muscles that are key to its terrifying bite strength.

▲ Brutus is a particularly large saltwater crocodile (*Crocodylus porosus*) found in and around the Adelaide River in Northern Territory, Australia. He (or she) is very recognizable thanks to a missing front limb.

Record-breakers

In 1979, a saltie was found drowned in a fishing net in Papua New Guinea, and was subsequently skinned. Its dried skin measured 20.3 ft (6.2 m) long, although the live animal would have been longer as the drying process would shrink the skin and also the tip of its tail was missing. A crocodile called Cassius at the Marineland Crocodile Park in Queensland, Australia, is generally considered to be the largest living saltie in captivity. He measured 17.3 ft (5.28 m) when he was captured in 1984, and at the time of writing is estimated to be 18 ft (5.48 m) long and weigh 2,200 lb (1,000 kg). These are only estimates because it is impossible to measure the crocodile when he is conscious and it is considered too dangerous to his health to try to sedate him.

A saltie called Gomek measured 17.8 ft (5.42 m) long and weighed 1,896 lb (860 kg) when he died at a zoo in Florida in 1997, but he was eclipsed by another individual called Lolong. That monster was an astonishing 20.2 ft (6.17 m) long and 2,370 lb (1,075 kg) in weight when he was sedated and measured in 2011

in a park in the Philippines, and might even have been a touch longer when he died two years later. Lolong was estimated to be fifty years old, but another saltie that died in a zoo in Russia in 1995 was thought to be at least eighty, as there was good evidence he arrived at the zoo in 1913 or 1915, and the owners of Cassius consider him to be 115.

Since trophy hunters target large individuals, and since the saltie population in northern Australia was reduced by 95 percent by hunting in the 1950s and 1960s, it has been feared that the genes conferring large size may have been lost from the population and no giants will occur again. However, I think this is unlikely for two reasons. First, saltwater crocs do occasionally undergo long-distance migrations, thereby refreshing the gene pool in northern Australia. Second, and more importantly, crocodiles grow throughout their life, so getting to large size is more about surviving to a ripe old age than about genes. With the decline in trophy hunting and the fall in favor of crocodile-skin handbags and shoes, we can hope to see even bigger salties sometime in the future.

Predictable Predators

There are at least 100,000 adult salties (and up to twice that number) along the rivers, deltas, and marshlands of northern Australia, but these are responsible for fewer than a handful of deaths per year. There are a number of reasons for this. First, not many people live in the region, those who do generally don't make a living in ways that require them to wade through swamps, locals know which swimming holes to avoid, and there is ample signage warning tourists of the danger. Second, the strongly territorial nature of the animals means that it is easier to predict where they might be (and might not be), compared to big sharks, for example, which have a nomadic lifestyle. Third, although salties can grow to a huge size if they live long enough, the average individual is relatively small and so any attack on a human is less likely to be fatal. Finally, crocodiles are not voracious eaters. They have a low metabolism and a very relaxed lifestyle, spending most of their time cruising around their territory, so attacks are often aimed at moving intruders along rather than eating them—even a big one might be satisfied giving you what it considers a little warning nip. That said, deliberate targeting of humans for food is known.

Although human deaths caused by salties are uncommon in Australia (around two per year), the risks in other parts of the species' range are harder to quantify. In places with high human populations where people must use river deltas for food and for work, and where infrastructure and government are sufficiently

remote, human deaths in the mouth of a saltie often go unreported. For this reason, it is definitely worth chatting to locals about salties before you enter the water in remote parts of Southeast Asia, especially at night, which is when salties do most of their hunting. If it's any comfort, it is highly unlikely that any saltie focuses entirely on humans for its dinner—the simple truth is that there is little these reptiles won't eat. It should be even more of a comfort that they do at least generally stay in the water, and they rarely attack humans (or anything else) on the land, so you can sleep in your hammock feeling relatively safe from saltwater crocodiles even if the river is within earshot.

▲ Measuring 20.2 ft (6.17 m) in length and weighing 2,370 lb (1,075 kg), Lolong was one of the largest salties before he died in captivity in a Philippines zoo in 2013.

◀ A life-sized model 28.2-ft (8.6-m) long of Australia's biggest known saltwater crocodile, which was shot in 1957.

Fearsome Freshies

The Nile crocodile is generally considered to be the second-largest crocodile after the saltie (see page 188), but there really isn't much in it, with the biggest Nile measuring more than 21 ft (6.4 m) long and weighing 2,400 lb (1,090 kg). The species is widespread across the waterbodies of central, eastern, and southern parts of Africa, and is responsible for substantially more human deaths than its saltwater cousin. South America has a giant freshwater croc too—but its future is far from assured.

The Deadliest Croc

Unlike salties, Nile crocodiles are gregarious and spend much of their time out of the water, dozing and basking on banks. They can hang out in groups because there is a clear pecking order, and when a chance for food does happen along, smaller individuals will generally defer to the larger ones without fuss. Their slow metabolism means they don't have to be constantly on the lookout for food, so not all individuals in an area will be hunting at any one time. In addition, Nile crocodiles have relatively small stomachs, so there are generally leftovers from a kill.

There are a number of reasons why the Nile crocodile causes more human deaths than any other. First, there are at least 250,000 Nile crocodiles and perhaps even twice that number. Second, they are found in a variety of waterbodies across a vast range that includes Somalia, Ethiopia, Kenya, Zambia, Tanzania, and Zimbabwe. Third, they live in parts of the world with very high human populations, where inhabitants rely on the same waterbodies used by the crocodiles for their drinking and washing water, transportation, and food gathering. Fourth, the Nile crocodile is more willing to come onto land than the saltie, and will pounce on prey from riverside undergrowth just as readily as launch itself from the water. Fifth, while salties are very territorial, and so more likely to attack humans just to move them on, Nile crocodiles are more likely to attack for food—consequently, a higher fraction of reported attacks by Nile crocodiles are fatal.

Last and by no means least, Nile crocodiles are very powerful predators. While they do not attack adult elephants, rhinoceroses, or hippopotamuses (*Hippopotamus amphibius*), they have been known to take adult giraffes (*Giraffa camelopardis*) and even healthy, fully grown lions (*Panthera leo*) drinking at the wrong part of the river at the wrong time. If a lion or hyena makes a kill near a river or lake, it is not uncommon for a crocodile to approach with a view to usurping the prey.

Nile crocodile
Crocodylus niloticus
Length: 20 ft (6.1 m)

Africa's Nile crocodile can grow to weigh more than 2,200 lb (1,000 kg). The biggest individuals have no fear of coming out of the water in search of prey, even when there are lions (*Panthera leo*) around.

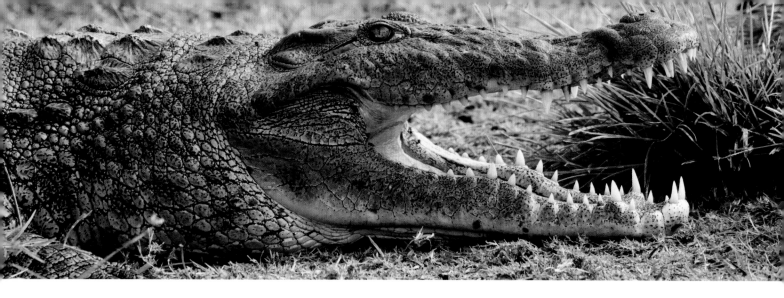

Critically Endangered

The third-largest living crocodile is the Orinoco crocodile, a critically endangered species that is mostly confined to the Orinoco River and its tributaries in Colombia and Venezuela. Males in excess of 20 ft (6.1 m) have been reported, but it is unlikely that any this size currently exist because the species has been extensively hunted for its skin. In theory, it is now protected, but in practice the countries involved do not have the resources to regulate a hunting ban across the remote and often inaccessible regions that make up its range. The future for the species does not look particularly bright.

Orinoco crocodile
Crocodylus intermedius
Length: 20 ft (6.1 m)

South America's Orinoco crocodile can also sometimes exceed 20 ft in length. In addition to pressure from humans, it faces competition from the spectacled caiman (*Caiman crocodilus*).

▲ The Nile crocodile doesn't have particularly terrifying teeth—it is the animal's size, its power, and its great bite force that make it such an effective killer.

Risky River

Scenes of Nile crocodiles gorging themselves on zebras and wildebeests in TV nature programs make terrifying viewing. The herd animals cross the Mara River on their annual migration across the Maasai Mara National Reserve in Kenya. This is a wide, fast-flowing river, often with steep banks, so TV crews and crocodiles both congregate at the times and places where they know the ungulates must cross. You sense that the potential prey know the crocs are there, but they also know that they must cross, and this pressure builds as more and more individuals arrive at the river. Eventually, some decide to hazard the crossing, at which point the water boils with crocs and 750 lb (350 kg) zebras are pulled under and held long enough for them to drown. Then the feasting begins. A zebra will keep even the biggest croc occupied for hours, so while some perish, most of the herbivores make it across the river intact and head for the lush grasslands on the other side.

Ancient Crocs

Crocodiles have been around for a very long time—they lived alongside the dinosaurs, and they somehow made it through the global catastrophe that finished off those giant reptiles. However, after this event the mammals diversified greatly and in particular started to take on roles requiring a larger body size, in turn marginalizing crocodiles to the role of aquatic ambusher. Today, we are finding some traces that suggest the ancestors of crocodiles led more diverse and active lives.

Size Estimates

Around the time of the dinosaurs, there were several lineages of very crocodile-like giants that are considered to be closely related to, but not direct ancestors of, modern crocodiles and alligators. The genera *Purussaurus*, *Machimosaurus*, *Sarcosuchus*, and *Deinosuchus* all seem to have included species that weighed an incredible 10 tons (9,000 kg) and were 30–40 ft (10–12 m) long, or twice the length of the very largest living crocodiles.

Modern-day crocodiles are very similar to each other in morphology and behavior, and the ancestors are comparable in anatomy to living crocodiles, so it seems safe to make inferences about these extinct giants on the basis of our understanding of living

species. For those species where we have only partial fossil remains, it is possible to estimate the full size of the body based on the body proportions of living crocodiles—and the figures indicate these really were giants. This is important in the context of a debate that has rumbled on in the scientific literature about whether ancient crocodilians were predators of dinosaurs. Several discoveries of dinosaur bones with teeth marks that fit these animals perfectly indicate that this is the case, but there is always the chance that the crocodilians simply scavenged on dead dinosaurs rather than killing them.

The very biggest crocodiles living today are apex predators willing and able to take large prey equivalent

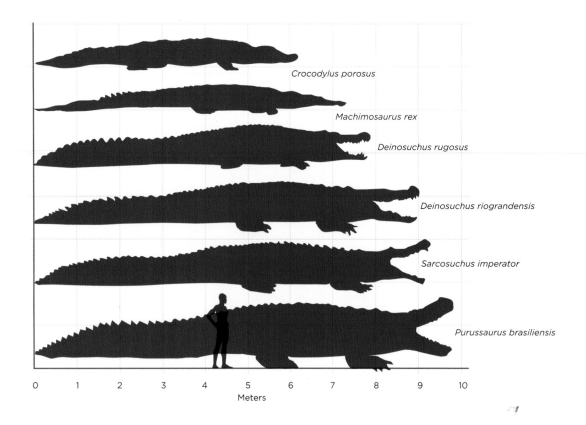

Crocodylus porosus

Machimosaurus rex

Deinosuchus rugosus

Deinosuchus riograndensis

Sarcosuchus imperator

Purussaurus brasiliensis

0   1   2   3   4   5   6   7   8   9   10
Meters

to a third of their body mass. This suggests that adults of some of the largest dinosaurs (like the sauropods) were probably safe from the ancient crocs, but anything up to 3.3 tons (3 tonnes) in weight could and would have been taken. *Deinosuchus* was around in North America at the same time as the tyrannosaurids, and it is not too fanciful to imagine a similar relationship between them as exists today between Nile crocodiles and lions in Africa.

Wider niches

Modern crocodiles stick very close to water, and many of them do all their feeding in this environment. We can see that their bodies have evolved to make them really good swimmers, even if that means they are more ungainly walkers. This is generally because predatory niches related to patrolling wide areas on land or running down prey have been filled by mammals. Instead, crocodiles have filled a gap as aquatic ambushers, a role that is less suited to mammals.

There were some mammals living at the time of the dinosaurs, but they were largely confined to niches requiring them to be small-bodied and nocturnal (which is still true of many mammals today, like rodents). This left a greater variety of predatory lifestyles open to crocodilians, and when we look at the fossils of early examples, we see that not all of these had the long, powerful tails and very short legs that are characteristic of crocodiles living today. While there certainly were some with these body plans, others walked a little more upright on longer

Sarcosuchus sp.
Weight: up to 8.8 tons (8 tonnes)

Sarcosuchus is a distant relative of today's crocodiles. Like many large crocodiles today, *Sarcosuchus* may have feasted on large animals that came down to rivers to drink or to cross while on migration. Their giant size would have conferred the ability to haul really large prey into the water and hold it under.

legs and had a less substantial tail. This suggests that they spent more time out of the water and likely hunted down their prey on land, making targeting of dinosaurs (especially juveniles) a lot more likely.

We see traces of this more terrestrial and perhaps more active past in today's crocodiles. When walking slowly, they generally adopt a sprawling gait that involves twisting their trunk every time they plant a foot. However, when they want to move a little faster, some crocodiles can rise up, straightening their legs and planting their feet directly beneath their body, making them a lot more agile. While mammals have a complex four-chambered heart that provides high levels of oxygen to their muscles when they are engaged in vigorous exercise, most reptiles have a simpler three-chambered heart that is sufficient for their more limited oxygen needs. Crocodiles, however, are the notable exception to this rule, having a four-chambered heart despite often engaging in a sedentary lifestyle. It seems likely that these features—the four-chambered heart and the ability to walk upright—are relics inherited from more active ancestors.

Amazing Amphibians

When we think about amphibians, frogs and toads generally come to mind, and although most are small, there are actually some pretty impressive ones. But amphibians also include the caecilians, which look like snakes, and the salamanders, which boast the biggest amphibian alive today.

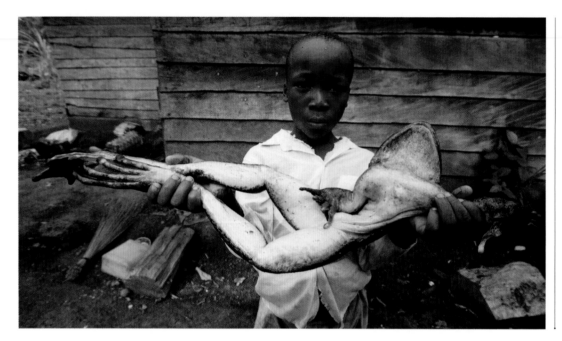

Goliath frogs
Conraua goliath
Weight: 13 lb (6 kg)

The large muscles in the hind legs of Goliath frogs make these amphibians attractive as food for humans across much of the species' range in Africa.

A Goliath Among Frogs

The goliath frog (*Conraua goliath*), from Cameroon and New Guinea, is the world's largest living frog—some individuals have a snout–vent length exceeding 1 ft (30 cm) and weigh more than 6.5 lb (3 kg). Scientists often characterize the size of the main body of an animal as the distance between the tip of its nose and its posterior orifice (the snout and the vent), to avoid values exaggerated by features like long tails, horns, and whiskers. In ecology, the goliath is a pretty typical frog, living in or close to rivers and eating a widely varied diet. It is becoming increasingly under threat as human populations within its range increase, given that it is sufficiently large to make an attractive meal and is generally a lot easier to catch than most other animals its size. Goliaths are also popular in zoos and as exotic pets, but as they don't seem to breed in captivity, collection of wild frogs for live export is putting a further strain on the population and the species should be made a conservation priority.

Another Awesome African

The African bullfrog (*Pyxicephalus adspersus*) is another impressive amphibian, with individuals commonly exceeding 4 lb (2 kg). It's much more common than the Goliath and is widely distributed across Angola, Kenya, Malawi, Mozambique, Namibia, South Africa, Swaziland, Tanzania, Zambia, and Zimbabwe. It is not as closely tied to rivers as the Goliath, and often lays its eggs in temporary pools left after heavy rains. If you do see one, take care not to get too close—African bullfrogs are voracious carnivores and one of very few frog species to have teeth.

Giant Salamanders

The Chinese giant salamander (*Andrias davidianus*) is the largest living amphibian, growing to 6 ft (1.8 m) in length and weighing nearly 130 lb (60 kg). Although it is technically classed as an amphibian, it never voluntarily comes out onto the land, and instead lives

African bullfrog
Pyxicephalus adspersus
Weight: 4 lb (2 kg)

Like most frogs, the African bullfrog will eat anything it can catch and swallow—no insect would be safe from the biggest individuals.

Chinese giant salamander
Andrias davidianus
Length: 6 ft (1.8 m)

Although the Chinese giant salamander is technically an amphibian, it is highly unusual for individuals to leave the water, which supports their big bulk.

permanently in fast-flowing rocky mountain streams and lakes in China. The wild population has declined dramatically in the last fifty years and continues to do so. The reason for this is simple economics: these animals are considered a luxury food item and can easily cost US$140–180 per pound (US$300–400 per kilogram) in restaurants. This financial incentive has triggered farming of the species in recent years as incomes in China have gone up, but so far this appears to be having a negative effect on wild populations since the industry is struggling with captive breeding and continually has to supplement stocks from the wild. The Chinese giant salamander's large size makes it very easy to hunt, and protecting the species is difficult given both the amount of hunting carried out and the remote rural areas involved. Habitat destruction associated with the rapid industrialization of China is another significant threat.

Frog or Toad?

There is no scientific distinction between a frog and a toad (which are grouped together in the order Anura), and the different common names have simply arisen through the vagaries of the English language. Toads are not necessarily more related to one another than they are to the frogs, so the naming is quite arbitrary. Generally, however, an amphibian is more likely to be known as a toad if it is bigger, has warty skin, is not brightly colored, has proportionately shorter hindlegs and so walks more than it hops, and is found in drier environments.

Changing with the Times

Between 370 million years ago and the dawn of the dinosaur era 240 million years ago, almost all vertebrates on land were a shifting balance of amphibians and reptiles of different types. None of these grew nearly as big as the dinosaurs, but it is worth checking out a few.

Spawn of the Devil

Frogs and toads have been around since the time of the dinosaurs. A particularly large specimen was found in 2008 in rocks dating back 70 million years. Media coverage of the discovery was particularly dramatic thanks to its scientific name, *Beelzebufo ampinga* (stemming from Beelzebub, or the Devil). It was dubbed by the press as the devil frog, devil toad, or simply the frog from hell, and artists depicted it chewing on a dinosaur. From this, you would expect it to be truly monstrous, but in fact its snout–vent length is estimated at just 9 in (23 cm), which although big for a frog is pretty much on a par with the largest species we see today. It is likely that the frog body plan puts a lot of strain on the bent hindlegs, limiting their size. As animals become larger on land, the general trend is for their legs to straighten and be held under the body (called a graviportal stance).

The Ascendance of the Amphibians

There has been life on Earth for more than 4 billion years, but for the great majority of that time it was confined to the seas and the land was essentially barren. Fast forward to 430 million years ago, and some plant life was beginning to emerge on land at water margins, and so inevitably invertebrate animals began to exploit this spread. By the early Carboniferous period 350 million years ago, there was abundant plant life across much of the land, this often taking the form of warm, wet rainforest. The land was also teeming with invertebrate life. Vertebrates began to emerge from the water to take advantage of all the invertebrate food, and at this point they were all amphibians.

Until relatively recently, scientists imagined that modern-day mudskippers were a good analogy for

Beelzebufo ampinga
Body length: 10 in (25 cm)

Beelzebufo ampinga was pretty big for a frog, with a body that was almost the size of a human head. It would have been big enough to swallow some hatchling dinosaurs whole.

the emergence of vertebrates from the water and the transition of some fish into amphibians. That is, they imagined that the impetus that allowed some fish to evolve the ability to survive for a time out of water was that they inhabited temporary pools that periodically became too small or dried up entirely, requiring the animal to move across the land in search of a new pool. However, too many of the fossilized footprints of the early walking fish that have been found were made underwater for this story to be completely feasible. By this time, the oceans were extensively populated with large predatory fish, and some prey fish had responded to this pressure by developing heavy armor as protection. However, this armor would have made swimming expensive in terms of energy, and so it seems that some armored fish took to walking along the sea bottom at least some of the time instead. Walking was probably a lot slower than swimming, but if you aren't investing as much energy in swimming then you don't need to find as much food to balance your energy budget.

This balance appears to have worked out for some fish, which adapted their fins progressively to allow faster and more energetically efficient walking. They adapted a pair of fins at the front of their body and a pair at the back into walking limbs, and it is from these fish that all land vertebrates evolved. (This is why modern terrestrial vertebrates all have four limbs arranged in two pairs, one near the head, and one set back at the other end of the body—except those

species that have lost their limbs through evolution, like snakes.) Some of these walking fish evolved to exploit invertebrate prey near the water's edge, and as their ability to breathe in air and support their weight on land increased, they became amphibians.

▲ Mudskippers comprise 32 species in the fish family Oxudercidae. They can breathe air and walk around on their pelvic fins, which allows them to feed on insects out of the water.

▼ Frogfishes (members of the family Antennariidae) are often very comfortable using their fins to help them walk along the seabed.

The Shift from Amphibians to Reptiles

For a few tens of millions of years, the amphibians were the only terrestrial vertebrates. In terms of diet, they flitted between exploiting insects and other arthropods on the land, and returning to the water to feast on fish and aquatic arthropods. This was the time of giant insects (see Chapter 6), so there was likely an incentive for some amphibians to become bigger in order to exploit the large prey on offer. Indeed, some of these amphibians grew to very substantial sizes—one was a powerful fish-eater 6–10 ft (2–3 m) long, whose niche today is filled by crocodilians like the Indian gharial (*Gavialis gangelicus*).

Although the Carboniferous period is described as a time when the land was dominated by warm, wet rainforest, not all the land would have been the same—just as we have variations today. In areas where rainfall was scarce, amphibians would have struggled to survive. The eggs of amphibians are more similar to those of fish than to those of reptiles and birds. Importantly, they don't have an impervious shell. Thus, unless an amphibian egg is laid in water or in a very damp substrate, it will soon dry out and the youngster inside will perish. Furthermore, amphibians can take in oxygen via their skin as well as breathing through their lungs; this is particularly handy when they are underwater, as it reduces the frequency with which they need to come up to the surface. However, skin adapted to absorb oxygen by diffusion can't have a

thick protective integument and must be kept moist (just as we need to keep our lungs moist). This means that adult amphibians also struggle in warm, dry conditions, because they lose water through their skin. They can thrive in warm weather, but only if there is a source of water close at hand. Being larger helps a little by reducing surface area compared to internal reserves of water, but even large amphibians are—and were—excluded from some areas of the world. As a result, some Carboniferous amphibians began to evolve more reptile-like traits, developing shells to their eggs and a thick, impermeable skin.

The Rise of the Reptiles

Near the end of the Carboniferous period, an event called the Carboniferous rainforest collapse occurred, when the Earth's extensive greenery was replaced by drier habitats and vast deserts. In the subsequent Permian period (298–251 million years ago), these conditions led to the dominance of reptile-like vertebrates over the amphibians. There would still have been some wet environments (for example, at the edge of the landmasses) and so the amphibians did not disappear entirely—after all, they are still with us now—but there was a great diversification of reptiles at this time. Some of these species became very big, and because there were more diverse feeding niches, some switched to herbivory and some to carnivory, preying on other vertebrates. Digesting plant matter

Inostrancevia sp.

Length: 11 ft (3.5 m)

For 50 million years before the rise of the dinosaurs, the largest animals on land were reptilian. They included this predatory *Inostrancevia* individual, which was long and had fearsome saber-tooth canines.

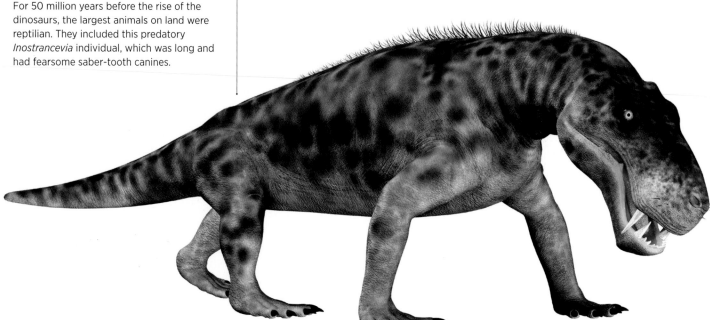

is easier if you are bigger, and larger carnivores can overcome larger prey. This great age of the reptiles continued until a sudden massive extinction event killed 70 percent of land organisms at the end of the Permian.

Members of the genus *Anteosaurus* were among the largest carnivores known from the Permian, and are estimated to have been up to 20 ft (6.1 m) long and 1,300 lb (600 kg) in weight. *Inostrancevia* species were sleeker-looking carnivores that approached the size of a modern rhinoceros at 11 ft (3.5 m) in length, but were armed with huge saber teeth. Pareiasaurs were a group of large reptilian herbivores, the largest examples of which were 10 ft (3 m) long and weighed 1,300 lb (600 kg).

So while there were large amphibians in the Carboniferous and large reptiles in the Permian, there was nothing to match the really spectacular sizes of the dinosaurs that came later, or indeed nothing to match the size of the largest mammals that arose after the dinosaurs. Bearing in mind that larger-bodied species have lower population densities, however, there might still be some truly massive beasts from these earlier eras waiting to be discovered.

Anteosaurus sp.
Length: 20 ft (6.1 m)

Towards the end of the Permian, one line of reptiles was evolving more mammal-like traits. Here, a predatory *Anteosaurus* is about to attack a plant-eating *Moschops*. There were true mammals by the time of the dinosaurs, but the dinosaurs generally outcompeted them and mammals remained scarce, small, and nocturnal until the dinosaurs became extinct.

▼ It is the hard shell of reptile eggs (like these turtle eggs) that give the group more freedom than amphibians in where they can lay their clutches.

Chapter 9
GREEN GIANTS

This chapter isn't just about plants, but about all primary producers, including both terrestrial and aquatic plants, and seaweeds and phytoplankton in the sea. What makes them primary producers is that they photosynthesize, capturing sunlight energy and using it to produce carbohydrates to fuel their growth and reproduction. Animals can't do this, and instead get their energy by eating plants, or by eating animals that eat plants, or animals that eat animals that eat plants—and so on. The primary producers are therefore the essential base of almost all food chains, and without them, there would quite simply be next to no life on the planet.

Top of the Tree

Ask most people what the biggest living thing on Earth is, and they would probably say a blue whale (*Balaenoptera musculus*). As we saw in Chapter 4, blue whales are massive—but they are nowhere near as big as the biggest trees. This is pretty good news in a way, because while blue whales are difficult to see, some of the largest trees in the world—the coast redwoods (*Sequoia sempervirens*) of California—are on show in various parks that are purposely designed to let viewers have a good look at them and take advantage of an impressive photo opportunity.

Tallest Living Tree

As you can imagine, there is a bit of competition between the different parks over which has the largest coast redwood. In fact, the exact site of the tallest—a tree named Hyperion—is kept secret to prevent sightseers damaging the local ecosystem, although it is somewhere in northern California. It stands 379 ft (115 m) tall, or about the height of a typical thirty-story high-rise; for comparison, the torch of the Statue of Liberty is a mere 300 ft (90 m) off the ground. Hyperion might be the current record-holder, but it is not a freak. There are hundreds of coast redwoods more than 300 ft (90 m) tall and at least thirty species of trees with examples in excess of 250 ft (80 m). There are historical accounts of trees that might have been taller than 379 ft (115 m), but no reliable evidence of any trees (even in the fossil record) reaching heights above 400 ft (120 m).

Coast redwood
Sequoia sempervirens
Height: up to 379 ft (115 m)

This is a stand of very tall coast redwoods from the Sequoia National Park in California. There are very few photos of Hyperion, the tallest of them all, to help keep its exact location a secret.

Giant sequoia
Sequoiadendron giganteum
Mass: to an estimated
1,200 tons (1,100 tonnes)

This is General Sherman, also growing in the Sequoia National Park and considered to be the world's biggest tree—not in height, but in volume and mass.

▲ The chance to walk through (or even drive a car through) a giant sequoia made the Wawona Tree a tourist attraction in Yosemite National Park from 1881 until it fell in 1969.

Largest Living Tree

As mentioned above, there are really immense individual trees that are easy to visit and admire. One is General Sherman, a giant sequoia (*Sequoiadendron giganteum*) growing in the Giant Forest of Sequoia National Park in Tulare County, California. It might only be 275 ft (84 m) tall, but it has a much thicker trunk than that of Hyperion (more than 102 ft/31 m, in circumference at the base), making it the largest known tree by volume of wood. Its trunk is estimated to have a volume of about 52,500 ft^3 (1,500 m^3) and to weigh about 1,200 tons (1,100 tonnes), equivalent to the weight of three Boeing 747s. To give you some idea of how thick the trunks of giant sequoias can be, there used to be another example at Yosemite

National Park called the Wawona Tree, through which park managers carved a road tunnel big enough fit a car. The tunnel was cut in 1881, and carriages and then cars drove though the still flourishing tree right until it collapsed in 1969. The tree fell under the weight of a heavy snowfall, and while it isn't known for certain whether the tunnel contributed to its demise, it is unlikely modern park managers will ever re-create such a tourist attraction. After all, the tree had experienced a heavy weight of snow plenty during its incredible 2,300-year lifespan. It doesn't feel quite right to modern values to increase visitor numbers and therefore revenue by taking an axe to such a venerable living organism.

The fossil record suggests that giant sequoias (also called giant redwoods) once had a much wider geographical range than their current distribution along the Pacific coast of North America, and specimens introduced in modern times to Europe, New Zealand, and temperate parts of Australia have generally thrived. In Scotland, for example, the Benmore Botanical Garden features an avenue of giant sequoias planted in 1863 and now featuring some examples close to 165 ft (50 m) tall. In France, the Disneyland Paris theme park has an impressive collection of the trees, and the country is home to the tallest example in Europe—an individual planted in Ribeauvillé in 1856 and now 260 ft (80 m) tall.

Weathering the Elements

Plants need light from the sun to fuel their metabolism and growth, and being shaded by neighbors is a major problem. The most obvious solution to this is to grow taller to escape their shade, but in doing so, those trees are shaded out in turn. From this, it is easy to see how such competition for light led to the evolution of some very tall plants indeed. But this upward growth comes at a cost. The trunk of a tree is essentially a support structure that lifts the energy-gathering leaves away from the shade of others; there is no advantage to it aside from this, and it is expensive for the tree—which is why not all trees grow to giant proportions.

A Weighty Problem

The most obvious limit to tree size is the strength required to support additional weight, as seen by the fact that the 2,300-year-old Wawona Tree fell while carrying the extra weight of a huge amount of snow on its foliage. As a child, you probably tried to build towers out of building blocks, but no matter how careful you were, they ultimately fell. The problem here is that if the tower is even slightly out of alignment, its center of gravity is not directly above the base. This produces a tendency to bend the tower in the direction of the lean, which in turn makes the tendency to bend even stronger—until eventually the tower topples. The tendency to bend increases with the height of the tower, and in the case of a tree you can imagine it forcing a trunk that is too tall to buckle and break somewhere along its length. What happens here is that the tree succumbs to its own weight, unless it is stiff enough to resist the bending force of that weight. The bigger the tree, the greater weight that has to be resisted.

Scientists have used calculations to estimate how big a tree would have to be before the tendency to bend and break becomes a problem. Depending on assumptions made about the exact shape of the tree, it seems that weight and bending force are not the factors that ultimately limit the height of a tree. Calculations indicated that thick trees could be anything from 650 ft (200 m) to 1,300 ft (400 m) tall before these issues become a serious problem, so clearly something else is limiting tree height.

Bending with the Wind

You might imagine that wind could be a factor limiting tree size. Winds push a tree from the side, the force of which has to be resisted by the weight of the tree and the soil holding its roots in the ground. However, as seen in the discussion on long levers and moment arms on page 19, it is clear that the force applied by wind will be felt most at the top of the tree where the leaves are, and that it must be resisted where the tree is anchored at its base. For tall trees, the main force of the wind is therefore quite a long way from the base of the tree, and this will apply a strong turning moment at the base.

It seems, however, that the risk of being uprooted or snapping in strong winds is not the factor that limits the height of the tallest trees. Part of the reason for this is that the incentive to grow tall is to avoid the shade of other trees (see page 206), so a tall tree is likely to be surrounded by other tall trees, and these neighbors will provide some protection from the wind. Another reason is that trees generally seem to invest more into their root system as they grow bigger above ground, such that the larger the tree, the better anchored it is to the ground. Snapping rather than uprooting is also unlikely, because the stiffness required to avoid buckling discussed above also offers sufficient resistance to being snapped by the wind.

◀ Trees often have strong religious and spiritual symbolism. This is a sacred fig (*Ficus religiosa*), a species under which Buddha is believed to have attained enlightenment and hence under which many people choose to meditate.

▲ Wind can have a huge effect on trees. On Pine Mountain in southern California, the strong winds—which are almost always from the same direction—have caused this tree to grow at an extraordinary angle, allowing it to minimize its exposure.

▶ It is easy to forget that half the volume of a tree can be out of sight underground. An Indian banyan (*Ficus benghalensis*) like this one begins life growing high up on another tree, so its impressive roots can be seen largely above ground.

The Significance of Water

We know that photosynthesis carried out in the leaves at the top of a tree requires water as well as carbon dioxide and sunshine. While a taller tree receives more sunshine because it is less likely to be shaded by its neighbors, it faces a greater challenge in delivering water from its roots to where it is needed in the leaves.

Surprisingly, the general consensus among most scientists is that the need to deliver water to the leaves is the ultimate factor limiting tree height. Water loss from the leaves is an inevitable consequence of photosynthesis. Plants have openings in their leaves called stomata, through which they absorb carbon dioxide, but inevitably water is lost via these openings. This water is replaced by water in the soil, which is absorbed by the roots and transported up the tree.

In the Pipeline

A plant's water-transport structure (called the xylem) resembles a set of long, narrow pipes leading from the roots all the way up to the leaves, and these pipes are filled with water. As water is lost from the leaves, this tends to draw water up the xylem to replace it. There is a force of cohesion to water, whereby its molecules stick together. This means that as some water molecules exit the leaves, they tend to drag the

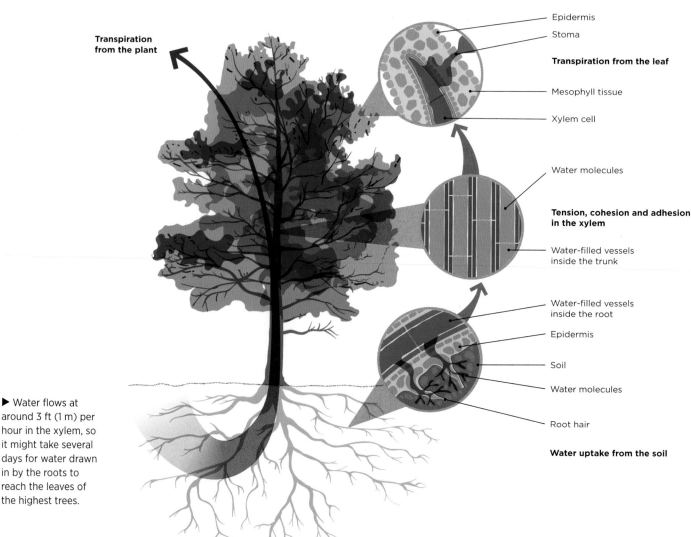

Transpiration from the plant

Epidermis
Stoma
Transpiration from the leaf
Mesophyll tissue
Xylem cell

Water molecules
Tension, cohesion and adhesion in the xylem
Water-filled vessels inside the trunk

Water-filled vessels inside the root
Epidermis
Soil
Water molecules
Root hair
Water uptake from the soil

▶ Water flows at around 3 ft (1 m) per hour in the xylem, so it might take several days for water drawn in by the roots to reach the leaves of the highest trees.

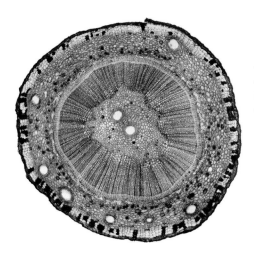

◀ In this artificially colored cross section of a stem, the round pipes of the xylem can clearly be seen.

molecules next to them toward the stomata to replace them. Inevitably, this process causes a chain reaction all the way down the plant that pulls water from the roots up the xylem to the leaves. Water diffuses into the roots from the surrounding soil to keep the whole xylem filled. This process is called transpiration and explains why all plants need water. In fact, plants need a lot of water. A medium-sized tree might go through something like 140,000 gallons (50,000 liters) of water a year, so a stand of twenty such trees would go through enough to fill an Olympic-sized swimming pool.

Opposition Forces

It is not difficult to see how delivering water this way becomes more difficult for a tree as its height increases. The transpiration pull has to work harder to lift water higher against the force of gravity, and also to overcome the frictional forces as the water is pulled up ever-longer pipes. Scientists have calculated that beyond heights of about 400–425 ft (120–130 m) it is virtually impossible for a tree to deliver sufficient water to meet its needs. From this, we would expect to find the tallest trees in places where there is high humidity at least some of the time, since this helps the trees: less water is lost from the leaves when the air is humid, and if the tree is parched, then water can travel in the opposite direction—from humid air into the leaves—to supplement water delivery from the roots. It now seems less surprising that many giant trees are found in California, where oceanic fogs often sweep in and can last for weeks at a time.

Although many giant fossil trees have been discovered, none appears to have been larger than the biggest trees living today. Indeed, it would be surprising if we found any taller than, say, 500 ft (150 m), as this would require an environment where the air was supersaturated with water.

Standing on the Shoulders of Giants

While the overwhelming majority of plants gather their water from the soil via their roots, there are some plants that get all their water from the atmosphere. Collectively, these are called epiphytes, and they are characterized by the fact that they grow on plants rather than in the soil. Imagine a plant that grows on one of the topmost branches of a tree in the tropical rainforest. Compared to a similar plant rooted in the soil, this epiphyte might receive a lot more light as there is less vegetation to block out its direct sunshine. After all, that is why rainforest trees grow so big. Unlike the trees, however, the epiphyte flourishes in the canopy without having to invest in a huge trunk—effectively, it takes advantage of the trunk of the tree it is sitting on. While epiphytes benefit from the height of trees without paying large structural costs, they do have to find an alternative source of water to that in the soil. For this reason, these plants are found almost exclusively in environments with moist air.

▼ One epiphyte can provide places for another's seeds to lodge, and thus a miniature forest of different types can become established on a single host tree.

Sensational Seeds

Plants have a simple structure compared to animals, basically comprising roots, shoots, and leaves, along with reproductive parts at certain times. But somehow the plant kingdom has managed to produce seemingly endless variations from this very basic template. Here, we look at these individual plant parts, starting with seeds, and ask why some are supersized.

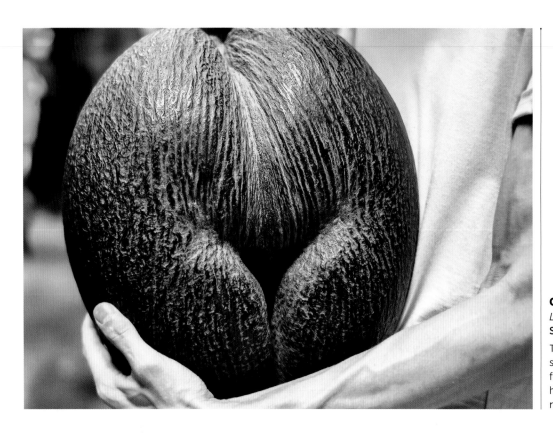

Coco de mer
Lodoicea maldivica
Seed weight: 37 lb (17 kg)

Traditionally, coco de mer seeds were hollowed out for use as bowls. Today, however, the species is very rare and closely protected.

Small vs. Large

Giant trees like coast redwoods take thousands of years to grow, but they start off from seeds no bigger than a typical sunflower seed. In terms of evolution, plants are no different from animals, their success being driven by the number of offspring they can produce that survive to produce offspring of their own. The smaller the seed, the cheaper it is in terms of energy for the plant to produce, and so the more seeds (and hence potential offspring) it can afford to make. This explains why seeds are often very small. However, there is a lot of variation between species in seed size—think about the difference between an avocado stone and an apple pip. It may pay some species to invest in a smaller number of larger seeds, if these larger seeds are more likely to grow into reproducing plants.

In contrast to small seeds, big seeds contain a larger energy store that will help the plant to grow in its earliest stages. Drawing on this store, it will be able to produce large leaves and a strong root system quickly, until these structures are sufficiently well developed that the plant can draw its energy from the sun and nutrients from the soil. The larger the seed store, therefore, the longer the new plant can last and the bigger it can grow before it has to be self-sufficient. It follows that big seeds might be particularly attractive in environments with poor soil, where the seedling has to establish a large root before it can acquire the nutrients it needs. If the seedling develops in the shade of a dense thicket of other plants, it will also need to grow tall and produce large leaves in order to compete for sunlight.

Coco de Mers and Coconuts

The reasons outlined above might well explain why the largest seed belongs to the coco de mer (*Lodoicea maldivica*), a palm tree that can be found only on two small islands in the Seychelles. The trees grow in a rocky, nutrient-poor soil of weathered granite, in dense stands where the coco de mer is virtually the only plant. Since mature trees grow 100 ft (30 m) tall and can have fan-shaped leaves measuring 30 ft (10 m) long and 13 ft (4 m) wide, seedlings have to grow quite tall and develop substantial leaves before they can gather enough sunlight to be self-sufficient. This is why the seed of this tree (itself also called a coco de mer, or sea coconut, double coconut, love nut, or Seychelles nut) can weigh up to 37 lb (17 kg)—about the same as five newborn babies.

Some people mistake the fruit of the coconut palm (*Cocos nucifera*) for a giant seed, but the actual seed can be seen if you look underneath the three indentations on the hard shell. Coconuts are commonplace in the tropics and subtropics because they readily float and, thanks to their tough outer shell, can survive being in saltwater for months or years without ill effect. However, coconuts also likely have humans to thank for their broad distribution. Seafarers have long seen the fruits as an ideal combination of food and water stores for long journeys. Coconuts don't rot easily, they float if accidently washed overboard and so can easily be retrieved, and they contain both plenty of nourishment and a fluid to slake thirst. Thus, it is likely that many early seafarers in tropical and subtropical regions took coconuts with them on their journeys. They may have deliberately planted coconuts in places, but some fruits will have washed overboard from vessels in rough seas, drifting far enough away that they were not recovered by the sailors, and colonizing new shorelines where they made landfall.

▲ Established coco de mer trees at the Vallée de Mai Nature Reserve on the island of Praslin in the Seychelles. It is clear here, in a situation where there is already intense competition for light, water, and nutrients from nearby established trees, why the plants need large seeds to help become established on relatively poor soil.

◄ Coconuts tend to get washed up on beaches. Sand is very low in the nutrients the plant needs to fuel its initial growth, so material is recycled from the fruit to help the seedling through the earliest sprouting stages.

Coconut palm
Cocos nucifera
Fruit weight: 3 lb (1.5 kg)

The coconut fruit can survive floating in the sea for months—the seeds inside have been recorded germinating even after the fruit has spent 110 days in saltwater.

Fabulous Fruits

One of the main functions of fruits is to offer a tasty treat to lure animals into consuming them and inadvertently dispersing the seeds they contain. We might expect plants to offer the largest fruits they can afford, since bigger meals will attract more and bigger potential seed dispersers.

Jackfruit

Artocarpus heterophyllus

Fruit weight: to 77 lb (35 kg)

The jackfruit has long been an important crop in Indian agriculture, with evidence of its cultivation dating back as far as 3,000 years ago.

◀ On falling to the ground, the jackfruit breaks open readily, releasing a distinctive odor. This attracts small mammals to feed on its sweet flesh, which incidentally ingest and then disperse its seeds.

Falling into Place

The granite islands of the Seychelles are (along with Madagascar) the oldest islands in the world, having broken away from India around the time the dinosaurs went extinct. In such small islands it is easy for plant and animal populations to die out, especially if they experience changing climates—as in the case of the Seychelles. Hence, prior to the arrival of humans the islands likely had very few tree species, which explains why the coco de mer (see page 211) dominates completely in the sites where it occurs.

When the fruit containing a large seed like the coco de mer falls, it might bounce and roll a little, but in general it will come to rest near the base of the parent, and that is where the seed has to germinate. From this, we can see why coco de mer forests are so dense. In this unusual situation, the best strategy for the parents isn't to produce lots of offspring, but instead to produce a small number of seeds with a food store large enough to allow them to thrive better than the offspring of near neighbors. In this respect, the reproductive strategy of the coco de mer is like that of humans—we also invest a great deal in a small number of offspring. In contrast, most plants take the opposite approach, producing numerous small seeds that they scatter widely in the hope that at least some will land in an ideal place for the seedling to flourish. However, such scattering is not attractive on tiny islands, where windborne seeds would be blown out to sea and there are few animals that could act as seed dispersers.

Jumbo Jackfruit and Colossal Cultivars

The coco de mer fruit can weigh as much as 65 lb (30 kg), but surprisingly, it is not quite the biggest fruit. The jackfruit (*Artocarpus heterophyllus*), also known as the jack tree, jakfruit, or just jack or jak, is a species of tree in the fig family (Moraceae). When mature, it can produce 100 fruit at once, each weighing up to 77 lb (35 kg). It grows widely across Asia and

Pumpkin
Squash (*Cucurbita* sp.) cultivar
Fruit weight: to 1,750 lb
(795 kg)

The pumpkin is the largest competitively grown vegetable. Here, at a competition in the USA, entries sit on commercial forklift transportation pallets waiting to be weighed.

▼ Until comparatively recently, giant ground sloths (*Megatherium* sp.) were widespread in South America, and several tree species appear to have evolved huge fruits in order to facilitate consumption, and thus seed dispersal, by these giant mammals.

the fruit are used in many cuisines throughout the region. In the wild, when the fruit falls it splits open to reveal small seeds distributed throughout the flesh, which is readily consumed by a range of mammals and birds. Because the seeds are small, some will be inadvertently swallowed by the frugivores, which later defecate them unharmed some distance from the parents, thus effecting seed dispersal.

Humans have grown fruit-bearing plants for thousands of years, and many cultivated fruits are now consequently larger than their wild cousins. The most extreme example of this is the pumpkin, a squash (*Cucurbita* species) cultivar. Growers, particularly in the USA, compete to display the largest pumpkins at festivals. The winning entries often weigh more than 1,300 lb (590 kg), and in some cases exceed 1,750 lb (795 kg)—easily big enough to make a coach for Cinderella!

Animal Attraction

The main function of fruit is to attract animals that might disperse the seeds inside. We have already discussed why some plants have big seeds (see page 210), but if they want these to be dispersed by animals, then only large animals will be up to the job. If a small monkey found a strawberry, it would happily swallow it in one go, seeds and all, but if it found an avocado, it would nibble the soft flesh and leave the seed behind. Thus, this small monkey is no use as a seed disperser for an avocado plant (*Persea americana*), which instead needs a large herbivore to ingest its large fruits. As far as avocados go, this is a bit of a mystery, since there are no animals

living in South America today (where the species originates) that are big enough to disperse the seeds.

So why do avocados have such large fruits and seeds? The likely explanation is that until a few thousand years ago there were giant ground sloths living in South America (see Chapter 4) that would have been big enough to eat them. The ground sloths are now extinct, but the avocado plants that relied on them for seed dispersal have managed to survive despite still producing fruit that has lost its purpose. This is probably because there are just enough alternative seed-dispersal mechanisms, such as landslips and flash floods. Like the giant sloths, many other large animals have gone extinct in the last 50,000 years, and there are plenty of examples from around the world of plants and animals that coevolved, only for the animal to die out—a phenomenon called evolutionary anachronism.

Fantastic Flowers and Luxuriant Leaves

Flowers are advertisements to attract pollinators, so we might expect these to be as big as the plant can afford. However, big blooms cost more energy and resources to make, so the plant has to trade off investment in flowers with investment elsewhere—such as in its roots and leaves.

Corpse lily
Rafflesia arnoldii
Flower diameter: 3 ft (1 m)

As its common name suggests, the corpse lily produces an exceptionally powerful odor very like that of rotting flesh, and in so doing attracts flies that carry pollen from one flower to the next.

Playing Dead

The role of flowers is generally to attract pollinators and advertise the availability of a nectar and/or pollen reward in return for pollination. Having a bigger advertisement potentially helps to draw more attention from pollinators. However, the very biggest flowers are not pronouncements of food availability, but instead achieve pollination by tricking insects rather than rewarding them. Some insects like to lay their eggs in the carcasses of dead animals, so that when their young hatch they can feed on the flesh. A few flowers take advantage of this by pretending to be dead animals in order to lure these insects. The insects may then pick up pollen on their body before realizing their mistake, or deposit pollen if they have recently been fooled by another plant of the same species. Carcass mimicry has evolved several times, and can be very convincing: the

flowers may smell like rotting flesh and be warm like rotting flesh, and they may be the color and texture of an animal (some are even hairy).

Full Bloom

Carcass mimicry is the trick played by the two plant species that produce the largest flowers, both found in the rainforests of Indonesia: the corpse lily (*Rafflesia arnoldii*), whose flower can be 3 ft (1 m) across; and the titan arum (*Amorphophallus titanium*), whose single bloom can exceed 10 ft (3 m) in height. Both of these flowers smell strongly of rotten meat. Part of the reason why they are so large is probably because at least some of their mimicry involves producing heat, just as in decaying meat. We know that having a large size helps with heat retention (see the discussion on surface area to volume ratio in Chapter 1), so a larger flower can more easily be significantly warmer than the foliage around it. Scientists have found that many insects that lay their eggs in dead animals prefer larger ones when offered the choice, probably because larger carcasses stay warm longer and offer a greater food source for the developing young. A large flower might also simply be more visible among all the other plants in the dimly lit rainforest.

Producing such large blooms requires considerable investment from the plant. The titan arum must grow for seven to ten years before it can produce a single bloom, and this lasts only a couple of days at most. We know a lot less about the growth of the corpse lily because it is not a conventional plant: it does not have leaves, stems, or even roots. Rather, it lives as a parasite inside a specific species of vine. It consists of thread-like strands that are entirely embedded within the vine and draw nutrients from its host. You could walk past affected vines and have no idea the corpse lily was there—unless it happened to be in flower, which is the only part of the plant that ever sees the light of day.

Titan arum
Amorphophallus titanium
Inflorescence height:
10 ft (3 m)

The titum arum produces the largest unbranched inflorescence (cluster of flowers). Like the corpse lily it produces an exceptionally strong smell of decaying flesh to attract pollinating flies.

Amazon waterlily
Victoria amazonica
Lily pad diameter: 10 ft (3 m)

Here, two sisters are easily supported on a single Amazon waterlily pad in a botanical garden in China. My guess, however, is that their combined weight is significantly less than mine.

Single Flower or Inflorescence?

Flowers are such varied things that it is hard to pick an absolute winner as the largest flower. While the corpse lily is probably the largest single flower and botanists describe the titan arum as having the largest unbranched inflorescence, the largest branched inflorescence in the world belongs to the Asian talipot palm (*Corypha umbraculifera*). An inflorescence is a group or cluster of flowers arranged on a stem that is composed of a main branch or a complicated arrangement of branches. Like the titan arum, the talipot palm also needs to grow for a long time before it can flower—in its case, thirty to eighty years. It produces a stunning 20–25 ft-long (6–8 m) inflorescence of millions of flowers, then, once some of these have formed into seed-bearing fruits, the tree dies.

Talipot palm
Corypha umbraculifera
Inflorescence length: 20–25 ft (6–8 m)

A talipot palm might wait 80 years before producing its stunning inflorescence, and then after those flowers develop into fruits, the tree dies.

Giant Vegetation

For a plant, larger leaves mean more photosynthesis, but they are expensive to build and maintain. Palm trees in general have huge leaves, and the winner here is probably the tropical African palm (*Raphia farinifera*), whose compound pinnate leaves can each measure 80 ft (25 m) in length. The mature leaves are so large, in fact, that they are often used in the construction of livestock fencing and as roofing thatch. The fibers of surface layers of immature leaves can be peeled away to produce raffia, which is used in the manufacture of a huge diversity of handicraft goods.

When it comes to large single leaves, the clear champion is the Amazon waterlily (*Victoria amazonica*), which produces lily pads 10 ft (3 m) across. These are so robust that they can support the weight of a human—as shown in the photo above. The species has been popular in Europe since Victorian times, where it has been cultivated in heated glasshouses.

Ocean Constraints

Before closing this chapter, we should consider why the world's oceans are nearly devoid of giant plants. At first sight, this seems a strange question to ask—after all, we are so used to the idea that plants grow on the land and not in the sea. But how can we understand this pattern? In the sea as on land, photosynthetic organisms make food by capturing energy from the sun, and animals then eat them or each other. In the sea, however, the photosynthesizers are not plants but phytoplankton. These include a whole range of organisms, most of which are the opposite of the coast redwoods, being too small to see with the naked eye.

Out of their Depth

Part of the reason why there are no large marine photosynthesizers is that water absorbs light much more effectively than air, so at depths below around 300 ft (100 m) it is too dark for photosynthesis to take place. Plant seeds have enough stored energy to grow a stem up through a few inches of soil to reach the sun, but no seed on an ocean floor thousands of feet below the surface could store enough energy for its seedling to grow up into the zone where it could start to gather energy from sunlight. This is why we generally see seaweeds (which are actually photosynthesizing algae rather than true plants) on the rocks very near shore, where the water is much more shallow and even a newly established seaweed receives enough sunshine to fuel its growth. That said, some seaweeds can be pretty huge. The giant kelp (*Macrocystis pyrifera*) is often described as the largest, growing more than 160 ft (50 m) long and at a rate as fast as 2 ft (60 cm) a day. It can form dense underwater forests that offer small fish somewhere to hide in the often featureless ocean. But these forests also hug the coasts, as the waters soon become too deep for the kelp to survive.

Giant kelp
Macrocystis pyrifera
Length: to 160 ft (50 m)

The giant kelp is not only the largest seaweed but is also one of the fastest-growing organisms, extending by as much as 2 ft (60 cm) in a single day. The kelp often grows in dense "forests". These forests can provide somewhere for marine species to hide from predators – or just somewhere to congregate – so are often biodiversity hotspots in the oceans.

Floating Free

Does a seaweed really need to be anchored to the seafloor, and are there any giant free-floating photosynthesizers? After all, seaweeds don't have a root system as in most plants, just a holdfast that anchors them to the rock but doesn't provide water and nutrients, which are instead absorbed from the surrounding sea water. In fact, two species of seaweed do have a completely free-floating existence, namely *Sargassum natans* and *S. fluitans*. Their names give a clue as to why this mode of life is so rare: they are both found in the Sargasso Sea. This is an area of the Atlantic famed for its almost continual calmness—early sailors had to avoid it because there are no winds or currents here, effectively trapping them. Because of these conditions, the Sargasso contains the so-called North Atlantic garbage patch, a vast accumulation of floating man-made debris, but it is also home to the free-floating seaweeds, which can thrive here for the years it takes them to complete their life cycle.

In other parts of the world's oceans, a free-floating seaweed would be carried along by the currents through waters that differ considerably in temperature, salinity, nutrient profile, and dissolved gases.

A seaweed that is adapted for one set of conditions would therefore soon find itself transported from its comfort zone into waters that didn't suit it at all. Phytoplankton are carried along on currents, too, but these tiny organisms can reproduce very quickly—in just days or even hours—and so can complete their life cycle and produce offspring before they are transported to conditions where they can't flourish. Some of their offspring will be subtly different from the parents in ways that allow them to flourish in these new waters, and so the tiny phytoplankton thrive in the seas and large photosynthesizers are very scarce indeed.

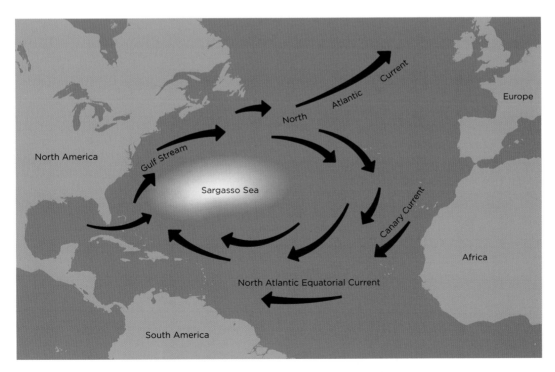

▲ Phytoplankton can grow so densely that they change the color of the sea—here to a lurid green shade. Simply by analyzing satellite photos to look at the color of the surface of the sea, scientists can measure plankton concentration over huge expanses of ocean.

◄ The Sargasso Sea in a large area of the Atlantic Ocean known for its very slow-moving water currents and winds. Free-floating seaweeds can live here because there are no currents to carry them to less suitable areas.

Final Thoughts
Being Organized Instead of Being Giant

We have seen throughout this book that many different lineages of plants and animals have evolved huge-bodied forms that capture the imagination with their majesty and power. But we should finish by reminding ourselves that the giants are the exceptions—fascinating exceptions, but exceptions nonetheless—and that most of the living matter around us flourishes on a smaller scale. We should also ask what sets humans apart, for we dominate the Earth like no other animal, alive or extinct, and yet we are not one of nature's true giants.

Most Life is on a Small Scale

I have hugely enjoyed reading and thinking about giants for this book. But giants are the exception—most life is very small. Scientists have tried to estimate how much biomass (living matter) there is in the world and have then divided this up into different groups. They describe such giant weights in terms of gigatonnes of carbon (Gt-C), carbon being a chemical element found in all living matter, and a gigatonne being 1 billion metric tonnes, or 1.1 billion tons. From these complicated calculations, it turns out that plants totally dominate animals (at 450 Gt-C, compared to 2 Gt-C), and that at 70 Gt-C even microscopic bacteria dominate animals. Looking more closely at the animals, half of the biomass of this group is made up of the arthropods (1 Gt-C, mostly insects), whereas wild mammals are trivial in comparison (0.007 Gt-C). And if we look at those wild mammals, two-thirds of the species are rodents and bats. This proves the point that even in the mammals—the lineage that has produced the massive blue whale (*Balaenoptera musculus*) and African bush elephant (*Loxodonta africanus*)—small body size is definitely the norm.

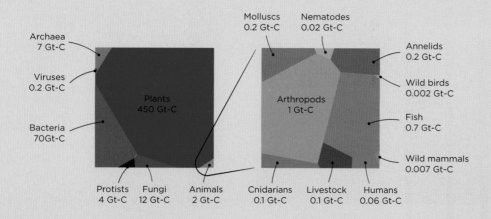

Molluscs 0.2 Gt-C

Nematodes 0.02 Gt-C

Archaea 7 Gt-C

Viruses 0.2 Gt-C

Bacteria 70Gt-C

Plants 450 Gt-C

Arthropods 1 Gt-C

Annelids 0.2 Gt-C

Wild birds 0.002 Gt-C

Fish 0.7 Gt-C

Wild mammals 0.007 Gt-C

Protists 4 Gt-C

Fungi 12 Gt-C

Animals 2 Gt-C

Cnidarians 0.1 Gt-C

Livestock 0.1 Gt-C

Humans 0.06 Gt-C

◀ Scientists Yinon Bar-On, Rob Phillips, and Ron Milo produced this estimate of how the mass of all living matter on Earth is divided between the different groups of organisms.

▲ Above: Killer whales (*Orca orcinus*) form tight social groups; coordinated attacks are often key to their success as predators.

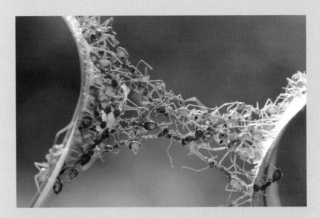

▲ Above: A single ant could not jump the distance between these two plant stems, but many ants working together can form a bridge between them out of their bodies.

Why is Small so Beautiful?

One reason why small animals dominate is that the same amount of resources can support a much larger number of, say, mice than deer. In addition, smaller-bodied animals generally have larger population sizes (and faster rates of reproduction), which probably makes them less vulnerable to extinction. Humans have driven a lot of species to extinction over the last 50,000 years, and this process is accelerating. Larger-sized species have suffered most, part of the reason for this being that they are more vulnerable (as we have just explained). However, it is also the case that large-bodied animals have been preferentially targeted because they are more obvious competitors for food, they are a more substantial meal in themselves, and/ or they are seen as a threat to us or our livestock. The last reason why smaller-bodied animals are more numerous is that they tend to be lower in the food chain and so have more food available to them (see Chapter 1). In addition, their small size allows them to package up the environment into different niches more efficiently. If you consider a grassland meadow, you can imagine one or two ways in which deer could make a living there—perhaps a small-bodied species preferentially targeting new shoots and a larger-bodied species consuming tougher vegetation. But there are many more ways you could imagine rodents coexisting in that meadow—some eating different plants, some eating insects, some eating a mix of the two, some active by day or night, some living out in the open or staying close to cover, some burrowing or not... the list goes on. As a generality, small-bodied species can coexist more easily than larger-bodied ones.

Getting Organized

Giant size confers immense power, but the other way to get power is through teamwork. Humans dominate the Earth now. We achieved this through teamwork, social organization, and culture. Similarly, while killer whales aren't the biggest predatory whale species, they are the most widely feared because their teamwork means they can target even the largest animals. Ants are tiny, but next to humans they are perhaps the most successful animals on the planet today—20 percent of all animal biomass comprises ants. Like humans, they are a supreme example of teamwork, building a communal place to live and being able to find much more food through working together. They do this despite having tiny brains, so to be successful, you don't have to be big and you don't have to be smart, but it does help a lot if you are social.

There might be a message for humanity in all this. We may currently be dominating the planet, but we are not doing so in a sustainable way. The only way for humans to be truly successful on the timescales we have been talking about in this book is to ratchet up our sociality a bit more and work together for a more equitable, harmonious, and sustainable future. We can but hope.

Further Reading

A great book for helping you understand how evolutionary pressures cause different species to vary in size is John Bonner's *Why Size Matters: From Bacteria to Blue Whales* (Princeton University Press, 2006).

For wonderfully written guides on how animals work in a more general sense, I recommend any and all books by the late Steven Vogel: check out his *Wikipedia* page (https://en.wikipedia.org/wiki/Steven_Vogel) for a list of works. And for understanding how physical processes impact on organisms, Mark Denny's *Air and Water: The Biology and Physics of Life's Media* (Princeton University Press, 2006) is an excellent read. Books by both of these scientific writers are inspiring and make complicated concepts beautifully clear.

With regard to scientific journal articles, these are becoming increasingly available for free download. Simply call up Google Scholar on your browser, type in a few words on any subject that interests you in the search bar, and then see what comes up. As a starter, a lovely overview of marine giants by Craig McClain and colleagues, "Sizing ocean giants: patterns of intraspecific size variation in marine megafauna", published in the open access journal *PeerJ* in 2015, can be downloaded free of charge from https://doi.org/10.7717/peerj.715.

Index

Page numbers in **bold** include illustrations.

Acanthacorydalis fruhstorferi see dobsonfly
Acrophyseter (an extinct whale) 92
Aepyornis see elephant birds
aerodynamics **105–6**, 108, 109, 111, 113
Africa 20, 42–3, 48–9, 52, 120, 127, 129, 145, 153, 192, 196
Alamosaurus (a sauropod) 24
Alaska 44, 59, 60, 68, 69, 95, 184
albatross **104**, 105, **106**, 159
   black-browed **105**
   wandering 102, **103**
Aldabra Atoll, Indian Ocean 183
Aldabrachelys gigantea see tortoise, Aldabra
   A. g. hololissa see tortoise, Seychelles giant
ammonites **165**
Amorphophallus titanium see arum, titan
amphibians **30**, 140, 171, 180
   evolution 198–200
   largest living **196–7**
   species number **126**
Amphicoelias (a sauropod) 25
anaconda, green **173**
Andrias davidianus see salamander, Chinese giant
Antarctica 78, 118
Anteosaurus sp. (an extinct reptile) **201**
Anthropornis nordenskjoldi see penguin, Nordenskjoeld's giant
ants 94, **139**, **219**
   bullet 130
   leafcutter **16**

Anura order 197
Apatosaurus see Brontosaurus
Apis cerana japonica see honeybee, Japanese
Aptenodytes forsteri see penguin, emperor
arachnids *see* spiders
arachnophobia 144
Arambourgiania (a pterosaur) 114
archaea **218**
Archelon sp. (a turtle) 186, **187**
Archispirostreptus gigas see millipede, giant African
Architeuthis dux see squid, giant
Arctic 58, 59, ,156, 158
Arctodus simus see bear, North American short-faced
Ardeotis kori see bustard, Kori
Argentavis magnificens (an extinct bird) **108–9**
Argentinosaurus (a titanosaur) **24**
armadillos **54**, **122**
arthropods
   aquatic **146–9**, **150–1**
   extinct **152**, **153**
   biomass **218**
   exoskeleton/molting 137, 145, 148, **149**, 151
   heaviest (living) **147**, **151**
   land, largest **146**, **150**
   land *vs* sea 137, 146
   longest legspan (living) **147**, **150–1**
   size limitations 146–8
   see also crustaceans; insects; spiders
Artocarpus heterophyllus see jackfruit
arum, titan 214, **215**
Attacus atlas see moth, Atlas

Australia 116, 118, 130, 144, 156, 175, 176, 179, 186, 188, 190, 191, 205
avocado 210, 213

bacteria **218**
Balaenoptera bonaerensis see whale, Antarctic minke
   B. brydei see whale, Bryde's
   B. musculus see whale, blue
   B. physalus see whale, fin
banyan, Indian **207**
bats 30, **110–11**, **113**, 118, 136, 144, 218
bears **17**, 18, 21, 28, 57, 90, **122**, 145, 148
   black 61
   brown (grizzly) 58–9, **60–1**, 62
   Eurasian cave 60, **61**
   North American short-faced 60
Beelzebufo ampinga (an extinct frog) **198**
beetles 126, 128, 143
   dung 57
   elephant **127**
   goliath 127
   titan 127
   water 130
Berardius bairdii see whale, Baird's beaked
biomass 79, 81, **218**, 219
birds
   breathing system 29, 110, **111**
   eggs, biggest 120, **121**
   elephant **120–1**
   evolution 136
   flight **14–16**, 37, **103–6**, 108, 110, 136
   flightless 69, 101, 108, **116–17**, **120–1**, **122–3**
   largest extinct **120–1**, **122–3**

largest extinct flying **104**, **108**
largest living **120**
largest living flying **102**, **103**
   size limitations 14–16, 103, 123
   species number **126**
   terror **122–3**
   see also individual species
Birgus latro see crab, coconut
bite force 34, 189
bivalves **162–4**
body size
   arthropod, limitations to 146–8
   birds, limitations to 14–16
   carnivore *vs* herbivore 31, 51
   and digestion efficiency 17, 26, 28, 29, 51
   dinosaur, benefits/drawbacks of large 25, 26–7, 28–9, 51
   dinosaur *vs* mammal 51
   distributions **30–1**
   fish *vs* whale 82
   fish *vs* arthropods 147–8
   insect, limitations to **134–5**, 136, 137
   and metabolic rate 14, **16–17**, 20–1, 26, 51, 81, 82
   and reproductive rate 42, 51, 219
   and swimming speed 81
   and walking efficiency **19**
bone
   densest mammal 66
   deposition/rings **33**
   lightweight 108, 112, 114, 115
Brachiosaurus altithorax (a sauropod) **24**, **25**
brains 73, 77, 135, 172, 177, 219
breathing systems *see* respiratory systems

Brontosaurus excelsus (a sauropod) **24**
Bruhathkayosaurus matleyi (a sauropod) 25
"Brutus" (a crocodile) 190
Brygmophyseter (an extinct whale) 92
bullfrog, African 196, **197**
bustard, Kori 102
butterflies 128, **137**, 140, 147
    Queen Alexandra's birdwing **128**, 129

caddisfly **141**
caecilians 196
Caerostris darwini see spider, Darwin's
    bark
camels 19, 26, **122**
Canis lupus see wolf
Caperea marginata see whale, pygmy
    right
Carboniferous period **132**, 134, **135**, **153**,
    198, 200, 201
    rainforest collapse 200
Carcharhinus leucas see shark, bull
    C. longimanus see shark, oceanic
        whitetip
Carcharocles megalodon see megalodon
Carcharodon carcharias see shark, great
    white
Caretta caretta see turtle, loggerhead
carnivores
    amphibian 196, **197**
    avian **118–19**, **123–4**
    dinosaur 31, **32–7**
    mammalian **58–61**, **62–5**
    reptilian (Permian) **200–1**
    size *vs* herbivores 31, 51
    see also individual species
"Cassius" (a crocodile) **190**
cassowary, southern **17**, **116**
Casuarius casuarius see cassowary,
    southern
caterpillars 21, 57, 127, 128
cave
    bears 61
    lions 65
    paintings 45, 61
    spiders 144
Cenozoic 30, 31, 134
centipedes **149**
Ceratotherium simum see rhinoceros,
    white
    C. s. cottoni 48–9
Cetorhinus maximus see shark, basking
Changuu, Indian Ocean 183
Chelonoidis see tortoise, Galápagos
    C. abingdonii see tortoise, Galápagos
        Pinta Island
circulatory systems
    leatherback turtle 185
    mammal *vs* reptile 195
    open *vs* closed 147–8
clam, giant **162–3**, 164
climate change 45, 50, 57, 79, 81, 82,
    154, 212
"Clyde" (a bear) 60
cockroach, giant burrowing 130
coco de mer **210–11**, 212
coconut palm (*Cocos nucifera*) **150**, **211**
condor
    Andean **104**, **109**
    California 102, **103**
Conraua goliath see frog, goliath
cooling mechanisms **17**, 29, 41, 86, **181**
coprolites 86
cormorant, spectacled/Pallas's 69
Corydalus cornutus see dobsonfly,
    eastern

Corypha umbraculifera see palm, talipot
crab
    coconut **146**, **150**
    giant spider **148**
    Japanese spider **147**, **150–1**, 152
Cretaceous period **134**, **135**, 136, 162, **163**
Cretaceous–Paleogene extinction 82, 97,
    115, 118, 165, 187
crocodiles 27, 53, 58, 62, 67, 71, 96, 175,
    178, **181**, 184, 187
    extinct **194–5**
    Nile 123, 188, **192–3**
    Orinoco 188
    saltwater **188–91**, **194**
Crocodylus intermedius see crocodile,
    Orinoco
    C. niloticus see crocodile, Nile
    C. porosus see crocodile, saltwater
crustaceans 81, 83, 146, **150–1**, 166
    see also crab; lobster
Cucurbita spp. *see* pumpkin
Cyanea capillata see jellyfish, lion's mane
Cygnus buccinator see swan, trumpeter
    C. olor see swan, mute

damselflies 129, **140**
Darwin, Charles 145, 162, 183
Deinacrida heteracantha see weta, Little
    Barrier giant
Deinosuchus (an extinct reptile) 194, 195
    D. rugosus 194
    D. riograndensis **194**
Dendroaspis polylepis see mamba,
    black **175**
dental batteries 29, 40, **41**, 44
Dermochelys coriacea see turtle,
    leatherback
Diceros bicornis see rhinoceros, black
diffusion 123, 132–3, 137, 200
digestive systems 59, 83, 91, 166, **167**, 168
    efficiency and body size 17, 26, 28, 29,
        51
    sperm whale 91
dimorphism, sexual 88, 126
Dinoponera gigantea (an ant) 130
Dinornis robustus see moa, giant
dinosaurs
    biggest land animal 23, **24–5**
    biggest predatory **32–7**
    major groups 31
    metabolic rate 29, 34
    respiratory system 28–9
    size benefits/drawbacks 25, 26–7,
        28–9, 51
    size distribution **30–1**
    size *vs* mammals 51
    see also sauropods
Diomedea exulans see albatross,
    wandering
Diplodocus carnegii (a sauropod) **24**
DNA, mammoth 44, 45
dobsonfly 129
    eastern **129**
dogs **17**, 61, 62, **122**
    African wild (hunting) 95, 120
dolphins 52, 66, 84, 88, 93, 94, 96, 104, 147
dormouse, edible 129
dragonflies 129, **132–3**, 140, 145
Dromaius novaehollandiae see emu
dugong (*Dugong dugon*) **66**, 67
Dytiscus latissimus (a water beetle) 130

eagle **105**
    bald **15**
    giant harpy 118
    Haast's **118–19**

earthworm
    African giant 166
    giant Gippsland 166, **167**
echolocation 88, 89, 159
ectothermy 98, 136, 146, 147
eggs
    amphibian 171, 200
    bird, biggest 120, **121**
    insect 57, 214
    reptile 171, 177, 185, **201**
    sauropod **29**, 31, 51
    shell 123
Elasmosaurus (a pliosauromorph) 99
elephant **17**, **18**, 19, 20, 21, **26**, 29, 36
    African bush 25, **40**, 163, 218
    African forest 40
    Asian 40
    Asian straight-tusked **47**
    biology **40–2**
    extinct ancestors **44–7**
    poaching **42–3**
elephant birds **120–1**
Elephas maximus see elephant, Asian
emu 116
endothermy 41, 82, 87, 97, 98, 136, 147,
    166, 180, 184
Enteroctopus dofleini see octopus, giant
    Pacific
epiphytes **209**
Eschrichtius robustus see whale, gray
Essex (a whaling ship) 89
Eunectes murinus see anaconda, green
Eurycnema goliath see stick insect,
    goliath
evolution
    amphibian 198–200
    bird/mammal 136
    reptile 82–3, 200–1
exoskeletons 137, 138, 145, 147, 148, 150,
    151, 152
extinction events
    Cretaceous–Paleogene 82, 97, 115, 118,
        165, 187
    Permian 201
eyes, largest 96

fat 16, 33, 67, 77, 88, 97, 174, 180, 185
feathers 34, 35, 108, 113, 136
feces 57, 86, 91
feeding
    bird 107, 109, 110, 118
    carousel (orcas) 95
    filter 72, 75, 80–1, 82, 83, 97
    lunge/skim (whales) 80–1
    and neck length 28, 40
    see also digestive systems
Ficus benghalensis see banyan, Indian
    F. religiosa see fig, sacred
fig, sacred **206**
fish
    biggest bony **86–7**
    biggest extinct 86, **87**
    biomass **218**
    size distribution 30
    size *vs* whales 82
    size *vs* arthropods 147–8
    species number **126**
    walking **199**
    see also sharks
fishing 61, 66, 73, 78, 95, 141, 150, 156,
    160, 190
flight
    bat 110, 136
    bird **14–16**, 37, **103–6**, 108, 110, 136
    insect 126, 129, 136, 139, **140**
    pterosaur **112–15**

fly
    Brazilian Mydas 129
    caddis *see* caddisfly
    dobson *see* dobsonfly
    timber 129
flightless birds 69, 101, 108, **116–17**, **120–1**,
    122–3
Flores Island, Indonesia **176**, 177
flowers **214–15**
flying fox
    Indian 111
    Lyle's **110**
food chains 20–1, 180, 203, 219
footprints, fossilized **27**, 113, **153**, 199
fossilization 31, 86, 144, 152, 157, 162
frigatebird **107**
frogs 113, 129, 141, 144
    bull *see* bullfrog
    extinct **198**
    goliath **196**
    and toads 197
fruit **211**, **212–13**
fungi 16, 60, 117, 126, **218**

Galápagos Islands 180, 183
Galeocerdo cuvier see shark, tiger
Gauromydas heros see fly, Brazilian
    Mydas
Gavialis gangeticus see gharial, Indian
"General Sherman" (largest living tree)
    205
Gerridae (pond skaters) **138–9**
gharial, Indian 200
gigantism
    island 126
    whale 81, 82
gills 72
giraffe (*Giraffa camelopardis*) 26, **28**,
    29, 192
gliding **105**, 106, 109, 110
Glis glis see dormouse, edible 129
glyptodons **54–5**, 57, **122**
Goliathus see beetles, goliath
"Gomek" (a crocodile) 190
gomphotheres 46
grasshopper, terrible *see* weta
"Great American interchange" **122**
growth, fastest 156, **216**
Gymnogyps californianus see condor,
    California
Gyps coprotheres see vulture, Cape

hadrosaurs (duck-billed dinosaurs) 24
Hainosaurus (a mosasaur) 98
Haliaeetus leucocephalus see eagle,
    bald
Haliphron atlanticus see octopus, seven-
    arm
Harpagornis moorei see eagle,
    Haast's
Harpia harpyja see eagle, harpy
Hatzegopteryx thambema (a pterosaur)
    114
Hawaii 21
heart, mammal *vs* reptile 195
Hemipepsis (a wasp) 130
hemoglobin 132
herbivores
    mammal **40–51**, **52–3**, **54–7**, **66–8**
    reptile **180**, **201**
    sauropod **26–9**, 51
    see also individual species
Heteropoda maxima see spider, giant
    huntsman
hippopotamus (*Hippopotamus*
    amphibius) 40, **52–3**, 77, 192

Homarus americanus see lobster, American
  H. gammarus see lobster, European
Homo erectus 177
  H. floresiensis **177**
  H. sapiens 11, 177 *see also* humans
honeybee, Japanese 131
hornet
  Asian giant 131
  Japanese giant **131**
humans
  attacks by animals 53, 59, 61, 62, 84, 118, 173, 175, 176–7, 188, 191, 192
  biomass **218**
  and demise of animals 42–3, 45, 53, 54, 57, 69, 72, 219
  early artforms 44, 45, 61
  see also Homo sapiens
hummingbirds **15**, 144
hunting, big game 43
  see also ivory poaching; whaling
Hydrodamalis gigas see sea cow, Steller's
"Hyperion" (tallest living tree) **204**

Icadyptes salasi (an extinct penguin) 93
ichthyosaurs **96–7**, 98
India 47, 188, 212
Indian Ocean 67, 150, 160, 183
inflorescences 215
Inostrancevia (an extinct reptile) **200**, 201
insects
  aquatic 130, **138–41**
  largest extinct **132–3**
  largest living 125, **127**
  largest living flying **128**, 129, **140**
  longest-bodied **130**
  pollination 214, 215
  respiratory system **134–5**
  size limitations **134–5**, 136, 137
  wing 129, 131, **132–3**, **134**, 138, **140**
insulation 34, 89, 185
intestinal worms 91, **166–8**
intestine, longest 91
invertebrate, biggest **158–9**
ivory poaching **42–3**, 53

jackfruit (jakfruit) **212**
Jaekelopterus see sea scorpions
jellyfish 87, **154–5**, 157, 185
  lion's mane 156, **157**, 168
  Nomura's **156**
Jesus bugs *see* pond skaters
"Jonathan" (a tortoise) **183**
Jurassic period 86, **135**

kelp, giant **216**
Komodo dragon **176–7**, 178, 180, 181

larvae 57, **127**, 129, 140, **141**
leaves, largest **215**
Leedsichthys (an extinct fish) 86, 87
lepidoptera **128–9**
Leptomithrax gaimardii see crab, giant spider
liger **62**
lily
  corpse 214, **215**
  water *see* waterlily
Lineus longissimus see worm, bootlace
lion 25, 42, 43, 53, **62**, **64–5**, 192, 195
  American 65
  Eurasian cave 65
Livyatan melvillei (an extinct whale) **92–3**

lizards **176–9**, 185, 200
lobster
  American **147**, **151**
  European 151
  molting **149**
Lodoicea maldivica see coco de mer
"Lolong" (a crocodile) 190, **191**
"Lonesome George" (a tortoise) **183**
Loxodonta africana see elephant, African bush
  L. cyclotis see elephant, African forest
lunging (feeding method) 80
lungs 17, 110, **111**, 132–3, 135, 178, 200
Lycaon pictus see dog, African wild

Machimosaurus (an extinct reptile) **194**
Macrocheira kaempferi see crab, Japanese spider
Macrocystis pyrifera see kelp, giant
Macropanesthia rhinoceros see cockroach, giant burrowing
Madagascar 120, 145, 212
Magnapaulia laticaudus (a hadrosaur) 24
mamba, black 175
mammals
  breathing system 110, **111**
  carnivorous **58–61**, **62–5**
  evolution 136
  herbivorous **40–51**, **52–3**, **54–7**, **66–8**
  largest land 47, **50–1**
  largest living land 40
  marine *vs* marine reptiles 82–3
  milk 77, 93
  species number **126**
  see also individual species
mammoth (*Mammuthus* spp.) **44–5**, 187
  Columbian (*M. columbi*) 44
  southern (*M. meridionalis*) 44
  steppe (*M. trogontherii*) 44
  woolly (*M. primigenius*) 44
Mammut see mastodons
manatee 66
  African 67
  Amazonian 67
  West Indian **67**, **68**
Manta birostris see ray, giant oceanic manta
marsupials 118, 179
mastodons **46–7**, 122
mayflies 138, 145
Medusozoa (Cnidaria subphylum) 154
Megachasma pelagios see shark, megamouth
Megalania prisca (an extinct lizard) **179**
Megalochelys atlas (an extinct tortoise) **186**
megalodon 86, **87**, 92
Megaloprepus caerulatus (a damselfly) 129, **140**
Meganeura (an extinct dragonfly) **132–3**
  M. monyi 140
Meganeuropsis permiana (a damselfly) 140
megapodes (Megapodiidae) 123
Megaptera novaeangliae see whale, humpback
Megascolides australis see earthworm, giant Gippsland
Megasoma elephas see beetles, elephant
Megatherium see sloth, ground
Meiolania (an extinct tortoise) 186, **187**
Mesonychoteuthis hamiltoni see squid, colossal

metabolism
  arthropod 146, 148
  crocodile 191, 192
  dinosaur 29, 34
  endothermy/ectothermy 98, 136, 146, 147, 180
  jellyfish/squid 154, 155, 159
  mammal *vs* reptile 180
  reptile 82, 180, 191, 192
  sirenian 66, 67
  and size 14, **16–17**, 20–1, 26, 51, 81, 82
  whale 81, 82, 83, 159
Microchaetus rappi see earthworm, African giant
migration 67, 81, **122**, 181, 190, 193, 195
millipedes 148, **149**
  giant African **152–3**
mimicry, carcass 214
Miocene epoch 46, 186
moa, giant **117**
Moby-Dick (Melville) 92
Moffett, Mark 125
Mola mola see sunfish, giant ocean **86–7**, 157
mollusks (Mollusca) 97, 141, **162–5**, 218
molting, arthropod **137**, 145, **148**, **149**, 151
Mongolarachne jurassica (a spider) 144
monitor lizard 185
  crocodile **178**
  water 178, **179**
Moraceae (fig family) 212
mosasaurs 98, **99**
Mosasaurus hoffmannii 98
Moschops (an extinct reptile) **201**
moth
  Atlas **128**, 129
  sloth 57
mudskippers **198–9**
muscles
  flight, bird 14–16, 103, 110, 111, 114, 116, 117, 118
  flight, insect 131, 134
  jaw 34, 92, **189**
  oxygen delivery 132, 135, 147–8
  power and size 14–16, 103
mutualism 57

nautilus **164–5**
Neanderthals 61
necks
  giraffe **19**, **28**
  and head size 19, 28, 40, 50
  insect 135
  plesiosaur 98–9
  pterosaur 113
  sauropod 24, **25**, 28–9, 40
nematodes 218
Nemopilema nomurai see jellyfish, Nomura's
Nephila komaci (a spider) 145
New Zealand 93, 117, 118, 120, 125, 126, 127, 156, 158, 159, 205
North America 32, 46, 60, 61, **122**, 151, 153, 195, 205
North Atlantic garbage patch 217

octopus 91, 151, 164, 165
  blue-ringed 160, **161**
  giant Pacific **160–1**
  seven-arm 160, **161**
Odonata 129, 140 *see also* damselflies; dragonflies
oldest terrestrial animal 183
orca (*Orcinus orca*) *see* whale, killer
ornithischians **30**, 31

Ornithoptera alexandrae see butterfly, Queen Alexandra's birdwing
ostrich 29, 116, **120**
overheating 17, 26, 35, 41, 50, 51, 52
Oxudercidae *see* mudskippers
oxygen
  absorption, amphibian 200
  absorption, mammal *vs* bird 111
  absorption and surface area 17
  concentration *vs* insect size **134**, **135**, 136, 153
  delivery, mammals *vs* reptiles 195
  delivery, open *vs* closed circulatory system 146–7, 148
  demand and temperature 146
  diffusion, egg 123
  diffusion 132–3, 137, 200
Oxyuranus microlepidotus see taipan, inland

Pachydyptes ponderosus see penguin, New Zealand giant
Pacific Ocean 67, **126**, 150, 160, 162
Palaeoloxodon namadicus see elephant, Asian straight-tusked
Paleogene period **135** *see also* Cretaceous–Paleogene extinction
palm
  coconut 150, **211**
  talipot **215**
  see also coco de mer
Panthera leo see lion
  P. l. atrox see lion, American
  P. spelaea see lion, Eurasian cave
  P. tigris see tiger
  P. t. tigris see tiger, Bengal
Pantophthalmus bellardi see fly, timber
Paraceratherium (an extinct rhino) **50–1**
Paraponera clavata see ant, bullet
Parapuzosia seppenradensis (an ammonite) 165
parasites, intestinal 91, **166–8**
parasitoids 131, 148
pareiasaurs 201
Patagotitan mayorum (a titanosaur) 24
pearl, clam **164**
Pelagornis sandersi (an extinct bird) **104**, 106–7
penguins 92, **93**
  emperor 93
  New Zealand giant 93
  Nordenskjoeld's giant 93
penis, whale **77**
Pepsis (a wasp) 130
Permian period **134**, **135**, 153, 200, 201
Persea americana see avocado
Phaethon aethereus see tropicbird, red-billed
Phalacrocorax perspicillatus see cormorant, spectacled/Pallas's
Phasmatodea (stick insects) **130**
phorusrhacids *see* birds, terror
photosynthesis 74, 79, 163, 187, 203, 208, 215, 217
Physeter macrocephalus see whale, sperm **88–91**, 158
phytoplankton 79, 154, 203, 216, **217**
plankton blooms 79, 154, **217**
Platyceramus platinus (an extinct mollusk) 162, **163**
plesiosauromorphs 98–9
plesiosaurs **98–9**
pliosauromorphs **98**, 99
Pliosaurus **98**, 99
polar waters 76, 79, 81, 82, 86, 146

pollination 214, 215
Polygonoporus giganticus (a tapeworm) 166, 168
pond skaters (Gerridae) **138**
proboscideans 47
productivity, polar waters 76, 79, 81
proteins 77, 79, 140
pteranodons (*Pteranodon* spp.) **112–13**, 114, 115
pterodactyl **113**
Pteropus giganteus see flying fox, Indian
P. *lylei see* flying fox, Lyle's
pterosaurs 30, 31, **35**, 101, **112–15**
Puertasaurus (a sauropod) 24
pumpkin (squash) **213**
Purussaurus brasiliens (an extinct reptile) **194**
python
Burmese (*Python bivittatus*) **173**
reticulated (*P. reticulatus*) **172, 173**
Pyxicephalus adspersus see bullfrog, African

Quetzalcoatlus northropi (a pterosaur) **114–15**

Rafflesia arnoldii see lily, corpse
rainforests 128, 129, 144, 198, 200, 209, 214
Raphia farinifera (a palm tree) 215
ray, giant oceanic manta 74, **75**
redwood
coast **204**, 210
giant *see* sequoia, giant
reefs, coral 98, **151**, 162
reproduction rate, and body size 42, 51, 219
reptiles
eggs 123, 171, **201**
evolution 82–3, 200–1
flying **112–15**
largest living 71, 188
marine, extinct **96–8**
marine *vs* marine mammals 82–3
metabolism 29, 82
size distribution 30
species number **126**
see also crocodiles; lizards; snakes; tortoise; turtles
respiratory systems
bird *vs* mammal 110, **111**
dinosaur 28–9
insect **134–5**
rhea, greater (*Rhea americana*) **116**
Rheiformes (order) 116
Rhincodon typus see shark, whale
rhinoceros 19, 21, 40, 192
black 48, **49**
extinct 47, **50–1**
Indian 48
white **48–9, 50**
rodents 16, **17**, 21, 27, 30, 60, 126, 218, 219

saber-tooth cats **65**, **122**
salamander, Chinese giant **196–7**
Sarcosuchus (an extinct reptile) 189, 194, **195**
S. imperator **194**
Sargasso Sea **217**
Sargassum fluitans (a seaweed) 217
S. natans 217
sauropodomorpha 30
sauropods
biology **26–9**, 31, 51
largest/tallest **24–5**

size distribution **30–1**
see also dinosaurs
Sauroposeidon (a sauropod) 25
Schmidt sting pain index 130
sea cow, Steller's **68, 69**
sea scorpion **152, 153**
seals 58, 59, **85**, 93, 95
elephant 84, 159
seaweed 203, **216**, 217
seed
dispersal 212, 213
largest **210–11**
sequoia, giant (*Sequoiadendron giganteum*) **205**
Sequoia sempervirens see redwood, coast
Seychelles 183, 211, 212
Shantungosaurus giganteus (a hadrosaur) 24
shark
basking **72–3**
bull 84
extinct 75, **86, 87**
great white **84–5**
megamouth 74
oceanic whitetip 84
size *vs* whales 82
sleeper 159
teeth 75, **87**
tiger 84
whale 72, **73**
Shonisaurus popularis (an ichthyosaur) **96**
S. sikanniensis 96
sirenians **66–8**
skaters, gerrid **138–9**
sloth, ground **56–7**, **122**, 213
Smilodon populator (a sabre-toothed cat) 65
snakes **172–5**, 181
feeding **83**
largest ever **174**
sea 98
soaring 37, 102, 103, 104, 105, 108, 109
sound production, sperm whale 88, 90
South America 46, 53, 54, 56, 65, 93, 108, 109, 116, 118, **122**, 129, 130, 144, 174, 187, 193, 213
spermaceti 88, 89, 93
spiders 11, 129, 137, 140
Darwin's bark **145**
fossil 144–5
giant huntsman **144**
goliath birdeater **144**
tarantula 130, **131**
webs 144, **145**
spiracles **134**
spiral shells **164–5**
squash (pumpkin) **213**
squid 91, 162, 164, 165, 187
colossal **158–9**
giant 159
stick insect, goliath **130**
stings 130, 131, 152, 154, 156
stomach
shark 84
whale 81, 90, 91, 158, 159, 192
stomata **208**, 209
stride length 19, 26
Struthio camelus see ostrich
Stupendemys geographicus (an extinct turtle) **186**, 187
Stygiomedusa gigantea (a jellyfish) 155
Styxosaurus snowii (a plesiosaur) **99**
"Sue" (a *Tyrannosaurus rex* skeleton) **32–3**

Sumatra, Indonesia 162, **176**
sunfish, giant ocean **86–7**, 157
swan
mute 102, **103**, 106
trumpeter 103
swimming, speed and body size 81
symbiosis 163

Taenia saginata see tapeworm, beef
tapeworm, beef **166–8**
tarantula (spider) 130, **131**
tarantula hawk (a wasp) 130, **131**
teeth
crocodile **193**
elephant/mammoth/mastodon 40, **44**, 46
hippopotamus **53**
ichthyosaur 96–7
saber **65**, 201
shark 75, 86, **87**
Tyrannosaurus rex **34**
whale 80, 90, **91**, 92, **93**
temperature
and arthropod body size 146
body 17, 33, 41, 82, 84, 147, 166, 180
and ocean layer mixing 154
tentacles
jellyfish 154, 155, 156, **157**, 168
squid 158, 159
Thalassarche melanophris see albatross, black-browed
Theraphosa blondi see spider, Goliath birdeater
theropods 30, 31
Thylacinus cynocephalus (a marsupial) 179
Thylacoleo carnifex (a marsupial) 179
Thysania agrippina see moth, white witch
tiger **62–3**, 64, 84, 181
Bengal 62
Titanis walleri (a terror bird) **122–3**
Titanoboa cerrejonensis (a snake) **174–5**
titanosaurs 24, **28–9**
Titanus giganteus see beetles, titan
toads 196, 197, 198
tongue, blue whale 77
tortoise
Aldabra 183
extinct 186
Galápagos **180**, **182**, 183
Galápagos Pinta Island 183
Seychelles giant 183
trachea 111, **134**, 135
transpiration **208–9**
trees
fossil 204, 209
largest living **205**
size limitations **206–7**, 208, 209
tallest living **204**
Triassic period 135
Triceratops (an ornithischian) 24, 31, **37**
Trichechus inunguis see manatee, Amazonian
T. *manatus see* manatee, West Indian
T. *senegalensis see* manatee, African
Tridacna gigas see clam, giant **162–3**
tropicbird, red-billed 107
trunk (proboscis) 26, 40, 41, 50
tsunamis 177, 183
turtles 71, 96, 175, **201**
extinct **186–7**
leatherback 157, **184–5**, 187
loggerhead 161

Tylosaurus (a mosasaur) 98, **99**
Tyrannosaurus rex 31, **32–3**, **34–5**, **36–7**, 189, 195

Uberabatitan (a titanosaur) **28–9**
Ursus americanus see bear, black
U. arctos see bear, brown
U. arctos middendorffi see bear, Kodiak
U. spelaeus see bear, Eurasian cave

Varanus komodoensis see Komodo dragon
V. priscus **179**
V. salvadorii see monitor, crocodile
V. salvator see monitor, water
venom 131, 160, 161, 172, 175, 177
Vespa mandarinia see hornet, Asian giant
V. m. japonica see hornet, Japanese giant
Victoria amazonica see waterlily, Amazon
viruses **218**
volcanism 183
Vultur gryphus see condor, Andean
vultures 37, **103**, 105, 109, 110
Cape **105**

wasps 130, **131**, 141
waterlily, Amazon 215
"Wawona Tree" **205**, 206
webs, spider 144, **145**
weightlifters (human) **18**
weta 126, 128
Little Barrier giant 125, **127**
whale
Antarctic minke **78**
Baird's beaked 88, **89**
baleen 51, 76, 77, 78, **80–3**, 86, 88, 93, 94
blue 16, 19, 39, **76–7**, 78–9, 156, 204, 218
Bryde's **82**
fin **80**
gray 77
humpback **83**
killer 78, 84, 90, **94–5**, 185, 219
pygmy right 80, **81**
size, why so big? 19
size *vs* fish 82
size *vs* marine reptiles 82–3
sperm **88–91**, 158, 159
sperm, macroraptorial **92–3**
teeth 80, 90, **91**, 92, **93**
whaling **78**, **79**, 80, 84, 88, 89, 90, 91, 95
wind
and flight **15**, 103, 104, 105, 106, 107, 109, 110
and tree growth **207**
wings
bird **14–16**, 37, **103–6**, 108, 110, 136
insect **128**, 129, 131, **132–3**, **134**, 138, **140**, 147
pterosaur **112–15**
wolf 21, 61, 62, 181
worm
bootlace 168, **169**
earth *see* earthworm
intestinal 91, **166–8**
tape *see* tapeworm

xylem **208**, 209

Zygophyseter (an extinct whale) 92

Acknowledgments

This was a hugely exciting project for me. I have written books before, but never one I imagined being read outside of universities, and I have never written without the safety net of coauthors. So, more than any previous book of mine, I am hugely indebted to those with much more experience in the publishing world than me.

Jacqui Sayers at Quarto took my initial idea and helped shape it into a much clearer vision of a book. She also assembled a crack team to help me develop that book. David Price-Goodfellow masterminded the whole operation, was an extraordinarily creative photo-finder and idea-generator, and developed the visual shape of the book. Luke Herriott was the gifted designer who actually delivered on David's vision. Susi Bailey was a dream copy editor: meticulous but also creative. Her ideas and gentle humor have hugely enhanced the whole text. Sherry Valentine was the perfect proofreader, saving me from many embarrassing errors.

At Yale, Jean Thomson Black and Michael Deneen provided the perfect interface between author and publisher, helping me feel really included in the process of bringing the book to the wider world. In addition, Jean found two exceptionally thorough and thoughtful readers who also were highly influential in honing the book's contents.

Get in Touch

Everyone makes mistakes. If you feel that anything you read here doesn't make sense or seems plain wrong, or you are truly horrified that I have overlooked your favorite giant, then do drop me an email at graeme. ruxton@st-andrews.ac.uk and I'll try to right any wrongs if ever the publishers can be persuaded to let me write a second edition. Thanks for giving this a book a go—I hope you enjoy it.

Picture Credits

(t = top, m = middle, b = bottom, l = left, r = right)

Alamy Stock Photo: 61t WILDLIFE GmbH; 67 Wolfgang Pölzer; 74 BIOSPHOTO; 77b National Geographic Image Collection; 79t blickwinkel; 81b Minden Pictures; 82 Doug Perrine; 84 Jeff Rotman; 89b Newscom; 95b & 109b Minden Pictures; 121b WILDLIFE GmbH; 127br Xinhua; 131t Rick & Nora Bowers; 132t Ian Hubball; 133 Science Photo Library; 138b ephotocorp; 144b Chris Howes/Wild Places Photography; 145b Andrew Mackay; 146 Minden Pictures; 147t Mark Conlin; 148 Nature Picture Library; 151t Leonid Serebrennikov; 153bl Minden Pictures; 153br Jack Barr; 157b WaterFrame; 161t Cultura RM; 161m BIOSPHOTO; 163br stockeurope; 167t Science History Images; 167bl age fotostock; 177 Sabena Jane Blackbird; 184 & 185t National Geographic Creative; 187b Mark Turner; 188 Rafael Ben-Ari; 191b Genevieve Vallee; 196 Cyril Ruoso/Minden Pictures; 198 Sergey Krasovskiy/Stocktrek; 216 WaterFrame.

Joe Borkowski: 91b.

CSIRO: 156b.

Dreamstime.com: 16l Danolsen; 18t Pariyawit Sukumpantanasarn; 42b Maggymeyer; 49t Ecophoto; 85b Sergey Uryadnikov; 87m W.scott Mcgill; 89m Michael Valos; 94t Bhalchandra Pujari; 97t Eugen Thome; 102 Ecophoto; 103bl Gallinagomedia; 105tl Andreanita; 105bl EtienneOutram; 106t Mobi68; 106b Chmelars; 107t Jonathan Amato; 107b Natador; 111t Geza Farkas; 113t Tinamou; 127bl Bonita Cheshier; 137t Isselee;

139 Shutter999; 150 Rafael Ben Ari; 163t Pniesen; 164t Tervolina88; 178t Jillyafah; 178b Det-anan Sunonethong; 183b Natursports; 191t Reynan Ignacio; 193t Hel080808; 204 Uros Ravbar; 205l Nickolay Stanev; 210 Irina Nekrasova; 211lt Giovanni Gagliardi; 211b Fenkie Sumolang; 212b Teguh Jati Prasetyo.

Getty Images: 156t Lucia Terui; 215bl STR/AFP.

Monterey Bay Aquarium Research Institute: 155b.

Nature Photographers Ltd: 136 Paul Sterry; 168 Steve Trewhella.

Science Photo Library: 46 Jaime Chirinos; 51 Mauricio Anton; 56b Jaime Chirinos; 96b Jaime Chirinos; 99t Jaime Chirinos; 99b Jaime Chirinos; 108t Jaime Chirinos; 119 Jaime Chirinos; 121t Jaime Chirinos; 123 Jaime Chirinos; 153 Jaime Chirinos; 158b Sinclair Stammers; 174 Jaime Chirinos; 186l Herve Conge, ISM; 187t Dorling Kindersley/UIG; 195 Jaime Chirinos; 201t John Sibbick; 128b Pascal Goetgheluck.

Shutterstock: 2–3 Catmando; 4–5 Gudkov Andrey; 8 Eric Isselee; 10 Krunja; 15t Wang LiQiang; 15b moosehenderson; 16r Eric Isselee; 17t Paopijit; 17br yuris; 18b gualtiero boffi; 19 HainaultPhoto; 20 William Booth; 25 Dotted Yeti; 26t Jez Bennett; 26b Jeff Grabert; 28 oxry; 28–29 Catmando; 31 AmeliAU; 32 Vlad G; 34 Puwadol Jaturawutthichai; 35t Herschel Hoffmeyer; 35b Elenarts; 36 Elenarts; 37t Warpaint; 37b Herschel Hoffmeyer; 41t Paul Hampton; 41m kgo3121; 41b kriangsakthongmoon; 42t Four Oaks; 43t Martina Wendt; 43b Michael Wick; 44b starmaro; 45t Pavel Masychev; 48 Alan

Jeffery; 49bl gualtiero boffi; 49br y ischte; 53t Dennis Jacobsen; 54t Andrea Izzotti; 57 Kristel Segeren; 58 Mario_Hoppmann; 59t Sylvie Bouchard; 59b Igor Batenev; 61b Horia Bogdan; 62 Akulinina; 63tr PhotocechCZ; 65tl Jez Bennett; 65tr Catmando; 65b Armand Grobler; 66 vkilikov; 69b Nicolas Primola; 73t wildestanimal; 75t magnusdeepbelow; 75b divedog; 79b Pylypenko; 81t Chase Dekker; 83tr Dmytro Pylypenko; 83bl Sebastian Kaulitzki; 87t Warpaint; 87b & 89t wildestanimal; 90 Bradberry; 92 Herschel Hoffmeyer; 94b Tory Kallman; 95t Andrey Visus; 109t Martin Mecnarowski; 110 kajornyot wildlife photography; 112b Dariush M; 114 Valentyna Chukhlyebova; 116t Ondrej Prosicky; 116m bmphotographer; 116br Matt Cornish; 118 IgorGolovniov; 120 Johan Swanepoel; 128t SanderMeertinsPhotography; 130 LukaKikina; 137b Mathisa; 138t iliuta goean; 147b Luca Santilli; 149t nicefishes; 149bl reptiles4all; 149br PetlinDmitry; 151b Arunee Rodloy; 155t Ethan Daniels; 157t Martin Prochazkacz; 160 Kondratuk Aleksei; 161b Gerald Robert Fischer; 165t Oksana Maksymova; 167br Juan Gaertner; 172b Yatra; 173bl tanoochai; 173br Heiko Kiera; 175t reptiles4all; 175b Warpaint; 176 Sergey Uryadnikov; 180 MyImages; 181b JMx Images; 182 FOTOGRIN; 185b Ramukanji; 190 Manon van Os; 192 David Havel; 193br Sergey Uryadnikov; 197t MZPHOTO.CZ; 199t aDam Wildlife; 200 Catmando; 201b Lynsey Allan; 206 PIXA; 207t By dlhca; 207b jeep2499; 209b Kaca Skokanova; 212t Dmytro Gilitukha; 213t Bob Pool; 214 Alexander Mazurkevich; 215t Isabelle OHara; 215br Snow At Night; 217t onsuda; 219l Karoline Cullen; 219r frank60; 91t Pavel Skopets

Nobu Tamura: 213b.

Wellcome Collection, London: 63b.

Wikimedia Commons: 27 Dan Lundberg; 29r Etemenanki3; 33t Sarah Werning; 45b Magyar Földrajzi Múzeum; 47l Honymand; 47r Royroydeb; 52 Nevit Dilmen; 53b cloudzilla; 55 H. Zell; 60 Yathin S Krishnappa; 63tl ArtMechanic; 68b Maharishi yogi; 73b NOAA Fisheries Service; 77t NOAA Photo Library; 78t Customs and Border Protection Service, Commonwealth of Australia; 78b US Library of Congress; 80 Aqqa Rosing-Asvid; 83br Dakota Lynch; 85t Fallows, C, Gallagher, AJ, & Hammerschlag, N; 91m NASA; 93t Hectonichus; 93b Wilhelm Kuhnert; 97t Ballista from Charmouth Heritage Coast Centre, Charmouth, England; 103t JJ Harrison; 103br PhilArmitage; 104t Jaime A. Headden; 115t H. Zell; 117t Auckland Museum; 127t Dinobass; 129 Dehaan; 131b Michael J. Trout; 132b Ghedoghedo; 140 Katja Schulz; 141t & 141m Hubert Duprat; 144t Didier Descouens; 145t&m Lalueza-Fox, C, Agnarsson, I, Kuntner, M & Blackledge, TA; 154 BuzzWoof; 155m Alexander Vasenin; 159 Oren Rozen; 162 Rick Hankinson; 163b Daderot; 165br Bjoertvedt; 169 Bruno C. Vellutini; 172t Stefan3345; 173t Dave Lonsdale; 179t Nur Hussein; 179b Steven G. Johnson; 181t Gianfranco Gori; 183t David Stanley; 189t Louis Jones; 189b fvanrenterghem; 193bl Fernando Flores; 197b Petr Hamernik; 199b Steve Childs; 209t Anatoly Mikhaltsov; 211m Wmpearl

Nathan Yuen: 21b.